T0271342

FERTIGATION

FERTIGATION
A Novel Method of Applying Crop Nutrients

P. Soman

Chief Agronomist
Jain Irrigation (Global)
Jalgaon, Maharashtra, India

CRC Press
Taylor & Francis Group
Boca Raton London New York

CRC Press is an imprint of the
Taylor & Francis Group, an **informa** business

NEW INDIA PUBLISHING AGENCY
New Delhi-110 034

First edition published 2022
by CRC Press
4 Park Square, Milton Park, Abingdon, Oxon, OX14 4RN

and by CRC Press
6000 Broken Sound Parkway NW, Suite 300, Boca Raton, FL 33487-2742

© 2022 New India Publishing Agency

CRC Press is an imprint of Taylor & Francis Group, an Informa business

Print edition not for sale in South Asia (India, Sri Lanka, Nepal, Bangladesh, Pakistan or Bhutan).

British Library Cataloguing-in-Publication Data
A catalogue record for this book is available from the British Library

Library of Congress Cataloging-in-Publication Data
A catalog record has been requested

ISBN: 978-1-032-15745-0 (hbk)
ISBN: 978-1-003-24552-0 (ebk)

DOI: 10.1201/9781003245520

Preface

Fertigation requires a thorough understanding of the science behind the technology to make it deliver the immense possibility it offers in crop production. Though the idea of fertigation existed from the times of solution culture, it did not receive the necessary attention from among plant nutritionists and agronomists when it reappeared in the context of micro irrigation.

Fertilizer application in field agriculture has also not developed as a precision technology. Recommendations of the quantum of fertilizers required for a crop, at least in India are not based on current varieties of the crops, nor have they anything to do with the growth rate and developmental changes occurring while a crop is managed by the grower. Most of the fertilizer recommendations are itself very old and efforts to make them relevant to the current growing conditions, soil status, crop variety and crops reaction to the environment etc. are very limited. It is even worse when growers follow traders' recommendations whose idea is to sell more the fertilizer they supply. Not only lower yields and very low fertilizer use efficiencies, but the deterioration of soil and water bodies are the results.

In this scenario, a good knowledge of fertigation; its principle, process and field level application is critical to get more crop per unit nutrient; higher fertilizer use efficiencies.

It is essential to confirm fertigation schedules at the field level by regular and repeatable agronomy trials in the field. Unfortunately, almost all the trials so far concentrated only on testing the total amount of fertilizer for a crop when fertigated vis-à-vis soil applied. While it is important to quantify the crop requirement of fertilizers, it is equally important to figure out the time a particular nutrient to be introduced in fertigation in order to get the best effects of nutrient on crop yield and quality. More research in this area is required before standardized fertigation plans (or programs) for many crops are designed and recommended to the growers.

It is also necessary to introduce Fertigation as a major subject at the University level or even earlier in the academic curricula.

In this book, I try to explain the basics and project the future course this technology may take in assisting precision crop production. A lot of the analysis is based on direct field experience of last 3 decades of extension and farmer support programs conducted by the author along with a team he developed, among small land holder farmers in several states in India and countries overseas. The author extends a very deep gratitude to Jain Irrigation Systems Ltd., a company which acted as the backbone of introducing micro irrigation and fertigation to Indian farmers with a very practical approach- Research, Develop, and Demonstrate in house and Train the farmers throughout the crop cycle, so that they ultimately become field masters of the technology. They saw, they were Taught, and they Applied their learning and they Won. This is continuing in the farm.

Author

Contents

Introduction

Plants require three factors for growth and reproduction: light, water, and nutrients. The third of these factors, managing crops to provide an optimum nutrient supply, is where some of the major differences among farming systems occur. These differences frequently are described as biological vs. chemical methods of maintaining soil fertility. This distinction is meaningful, but the categories are not mutually exclusive. Farmers understand that nutrients are obtained by plants from both disintegration of biomass (i.e composting) and from additions of chemical compounds called fertilizers.

In both cases, nutrients are chemical elements in absorbable form. Plant roots absorb nutrients as inorganic chemicals dissolved in water. Farmers generally do not fully understand the whole process but they know that after fertilizer application the crop has to be irrigated.

Biological materials like manure are major nutrient sources on many "conventional" farms, as well as organic farms, while inorganic minerals (chemical materials) are compounded as chemical fertilizers and in water medium they give away nutrients in absorbable form. For example, fertilizer Urea yields NO_2, the absorbable form of Nitrogen when reacts with soil moisture.

This is the key understanding; that chemical fertilizers can convert to absorbable form only if water is available in the soil.

In fertigation the dissolving of fertilizers in irrigation water and then applying to the crop is all done as one step operation. So the farmers will come to understand slowly that the usual ionization (conversion of the fertilizer chemical into plant/ root absorbable form of mineral) needs water as a medium. Once they get this basic knowledge, they will practice fertigation. The incentives to do so are;

1) The whole operation of fertigation and irrigation are just one step only.

2) The results of following a well -designed fertigation program (*what, when and how much of fertilizer*) are higher yields and income.

3) As their understanding deepens they would also learn that for every gram of fertilizer applied thru fertigation they are getting higher crop yield and income.

How a farmer understands Plant nutrition and Fertigation?

What is that the farmer wants to achieve while he nourishes the crop with costly fertilizer inputs? He would like to get maximum output (yield and quality) of the produce; and for the maximization of output, fertilizer input should be utilized by the crop to its maximum potential. Thus it is not higher fertilizer dose that decides the highest yield; but the highest efficiency in crop's use of those fertilizers. The efficiency, yield/unit fertilizer, can be enhanced by the following factors;

a) Plant roots absorbing (in –take) total quanta of applied fertilizer and minimizing the loss of applied nutrients by not having spare quantities lying around the soil profile. This is possible only if the quantum of application matches with the quantum of need/root absorption.

b) Plants utilizing the absorbed nutrients for metabolic activities concurrent to absorption. This is possible only if the fertilizers are applied in times of need of the plants and not before or after. This compels the grower to know the physiological need of the nutrient during each developmental stages; which nutrient is required and when.

Let us see the uptake as function of growth. Famer can easily understand from the following figures the quantum of N, P or K uptake by Tomato (an example) as it grows to the season end.

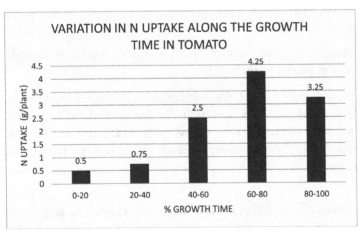

Figure. Variation in Nitrogen Uptake along the growth of field Tomato**. (**Data sourced from Fertigation by Kafkafi and Trachitsky 2011, and interpreted)

Tomato in the field takes up only 500 mg of Nitrogen during the first 20% of growth time (duration). Similarly it takes up only 4250 mg of Nitrogen during the 60-80% of growth time (approximately equivalent to 91-120 days period of growth). Suppose a farmer puts all N at land preparation and flood irrigates the field, as commonly done, tomato then will be subject to Nitrogen deficiency during the 60-80 % growth time when phenologically, the plant is producing flowers and fruits continually. That is why in Precision farming we recommend fertigation schedule prepared based on scientific knowledge of phenology of the crop.

Like that for Nitrogen, information is available for P, K, and other minor nutrients.

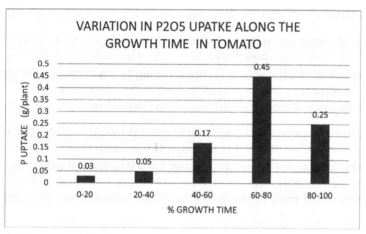

(**Data sourced from Fertigation by Kafkafi and Trachitsky, 2011 and interpreted)

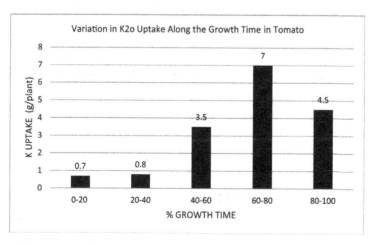

(**Data sourced from Fertigation by Kafkafi and Trachitsky, 2011 and interpreted)

Here the grower must have a powerful method to provide nutrients to the plants strictly following the predetermined schedule. Fertigation offers an opportunity to practice nutrient inputs: the grower is able to correctly decide the dose, type and time of providing the correct nutrient in precise quantity and at the required stage of plant/ crop development.

Thus the purpose of Agriculture, to maximize yield and quality of crops and minimize the costs of production while maintaining sustainability can be put into practice by fertigation. Two most important prerequisites to achieve all of these are the optimal and balanced supply of water and nutrients. Tuning of plant nutrient supply with the uptake and utilization rates by crops is essentially an art and is made doable by fertigation technology.

What is fertigation?

Fertigation' is the technique of supplying dissolved fertilizer to crops through an irrigation system. When combined with an efficient irrigation system (like drip method of irrigation) both nutrients (type and quantity) and water (volume or rate) can be manipulated and managed to obtain the maximum possible yield of marketable product from a given quantity of these inputs.

Often, solid fertilizer side-dressings are timed to suit management constraints (Grower convenience) rather than the physiological requirements of the crop. For example, most farmers will have experienced the dilemma of spreading fertilizer the day before heavy rain and then wondering how much of the fertilizer is either washed from the crop in run-off or leached below the root zone.

Continuous applications of soluble nutrients in very small doses overcome these problems, save labour, reduce compaction in the field, result in the fertilizer being placed around the plant roots (rhizosphere) uniformly and allow for rapid uptake of nutrients by the plant. To capitalize on these benefits, particular care should be taken in selecting fertilizers and injection equipment as well as in the management and maintenance of the irrigation system.

Fertigation requires water soluble fertilizers. The fact that some of the conventional fertilizers (Urea, Potassium sulphate, Ammonium sulphate etc.) are water soluble helped in introducing fertigation practice even if high cost specialized water soluble fertilizers are not available or not affordable to the growers.

Factors affecting his decision making on fertilizer application

First and foremost factor is availability of funds. And then the perceived benefit.

Farmers apply fertilizer only when they realize the benefits. Similarly they practice fertigation if only they get high yield and incomes.

Looking at the fertilizer use data of last several years it is also clear that, farmers here do not use balanced fertilizer program and the quantum of fertilizer types used over the years are skewed heavily to more nitrogen. Urea seems to be the only fertilizer most farmers buy and apply to the crops. The soils have become highly deficient in P and K and a whole lot of micro and secondary nutrients.

Persistent extension classes and visit to well grown and well maintained fertigation plots of crops are required to convince the farmers about the role of fertigation and balanced nutrition in crop production.

Farmers should be educated not only about the role of fertigation but the application schedule.

Crop uptake of nutrients are at a very low rate compared to the traditional application of fertilizers to the soil. This aspect of dosage has two main components: 1. The quantum of the dosage, and 2. The timing of application of each nutrient. Both these are intimately tied up with the crop growth and development stages. In traditional application of fertilizer, the importance of growth stages and development phases are not taken into account in a manner they should be.

Farmers either apply all fertilizers in full quanta at the beginning (along with land preparation) or at the most, split it into 2 or 3 parcels and apply at 2 or 3 interval, the partitioning in most cases have nothing to do with crop growth stages but more to do with the convenience of the farmer or some traditional calendar. So much so that applying fertilizer after the flower emergence in Banana or Rice or Wheat is a "NO GO" area in traditional agriculture; where as it is scientifically required for almost all crops to have maximum potassium to be taken up after flower appearance for fruit /grain development. Because these modern knowledge is not inculcated into the farmer support services of the public system an ordinary farmer is not aware of such basic things.

Because of all these system deficiencies in conventional fertilizer application guidelines, once fertigation is introduced and scientifically designed and tested schedules are made available to farmers a difficult time ensues during which farmers mindset about fertilizer use have to be changed.

Why fertilizer doses should be very small and more frequent

(i)   Application of fertilizer in small doses spread across the entire growing season in an effort to match the crop nutrient requirements, to improve nutrient uptake efficiency, minimize losses, thus to maximize the returns per unit amount of fertilizer; and

(ii)   Minimize nutrient leaching below the root-zone, particularly of nitrate form of nitrogen, which can have negative impact on raising its concentration in the groundwater above the maximum contaminant limit that is recommended for drinking water quality. In the case of large spacing planted tree crops, drip or under-the-tree micro sprinklers (micro irrigation) provide an opportunity to irrigate a certain portion of the total planted area, thus contribute to increased water uptake efficiency.

This book explains briefly, a number of basic issues on plant nutrition, soil chemistry, fertilizers, fertigation equipments and finally the practice of fertigation. There is a special chapter on fertigation scheduling- "how much of nutrients and when to apply". The book would help Agronomists and Extension officers to help the farmers to understand and practice fertigation besides helping students of agronomy and plant nutrition to understand the science.

1

Need for Precision Application of Nutrients to the Root

What is that the grower wants to achieve while he nourishes the crop with costly fertilizer inputs? He would like to get maximum output (yield and quality) of the produce; and for the maximization of out-put the fertilizer input should be utilized by the crop to its maximum potential. Thus it is not higher fertilizer dose that decides the highest yield; but the highest efficiency in crop's use of those fertilizers. The efficiency, yield/unit fertilizer can be enhanced by the following factors;

a) Plant roots absorbing (in–take) total quanta of applied fertilizer and minimizing the loss of applied nutrients by not having spare quantities lying around the soil profile. This is possible only if the quantum of application matches with the quantum of need.

b) Plants utilizing the absorbed nutrients for metabolic activities concurrent to absorption. This is possible only if the fertilizers are applied in times of need of the plants and not before or after. This compels the grower to know the physiological need of the nutrient during each developmental stages; which nutrient required when.

Here the grower must have a powerful tool or method to provide nutrients to the plants strictly following the above two factors. Fertigation offers an opportunity to practice nutrient inputs based on the above two factors: the grower is able to correctly decide the dose, type and time of providing the correct nutrient at precise quantity and at the required stage of plant/ crop development.

Thus the purpose of Agriculture, to maximize yield and quality of crops and minimize the costs of production while maintaining sustainability can be put into practice by fertigation. Two most important prerequisites to achieve all of these are the optimal and balanced supply of water and nutrients. Tuning of plant nutrient supply with the uptake and utilization rates by crops is essentially an art and is made doable by fertigation technology.

Fertigation' is the technique of supplying dissolved fertilizer to crops through an irrigation system. When combined with an efficient irrigation system (like drip method of irrigation) both nutrients (type and quantity) and water (volume or rate) can be manipulated and managed to obtain the maximum possible yield of marketable product from a given quantity of these inputs.

Often, solid fertilizer side-dressings are timed to suit management constraints (Grower convenience) rather than the physiological requirements of the crop. For example, most farmers will have experienced the dilemma of spreading fertilizer the day before heavy rain and then wondering how much of the fertilizer is either washed from the crop in run-off or leached below the root zone.

Continuous applications of soluble nutrients in very small doses overcome these problems, save labour, reduce compaction in the field, result in the fertilizer being placed around the plant roots (rhizosphere) uniformly and allow for rapid uptake of nutrients by the plant. To capitalize on these benefits, particular care should be taken in selecting fertilizers and injection equipment as well as in the management and maintenance of the irrigation system.

Fertigation requires water soluble fertilizers. The fact that some of the conventional fertilizers (Urea, Potassium sulphate, Ammonium sulphate etc.) are water soluble helped in introducing fertigation practice even if high cost specialized water soluble fertilizers are not available or not affordable to the growers.

2

Basics of Plant Nutrition

Only a very brief mention of various basic aspects of plant nutrition is given here. Associates wanting to know more details should refer to any one of the standard Plant Nutrition publication (PRINCIPLES OF PLANT NUTRITION by K. Mengel and E.A. Kirkby).

Increased production of food and cash crops is the need of our times; as land and irrigation water are limited and beyond any means of expansion. Crop yields per unit input should increase. Crop yield under natural production systems is limited by the availability of nutrients in the soil. Unless the supply of nutrients can be enhanced yields cannot be high.

Fertilizers perform the following functions

They supplement the natural soil nutrient supply, thus satisfying the requirement of crops for high yield potential.

They replace the nutrients removed at harvest, and also losses due to leaching, denitrification etc.

They help to improve poor soils and maintain good soil conditions under intensive cropping.

They help achieve sustainability.

Essential Elements

Plants require adequate supplies of 16 essential elements for good growth. The elements considered essential for plant nutrition are;

C, H, O –Major Elements non mineral come from CO_2 & H_2O

N, P, K, - Macro elements

Mg, Ca, S – Secondary Microelements

Fe, Zn, Mn, Mo, Cu, B Cl,–Microelements

Na, Ni, Si, Co- essential for some higher plants

13 of these elements are minerals which can be supplied or supplemented through the use of fertilizer, while the other 3, carbon, hydrogen and oxygen are absorbed either from the atmosphere as gases or in solution from the soil. The 13 essential mineral elements are all of equal importance to plant health, but are required in varying quantities. The exact quantities of each element required depend on the plant species and stage of growth but they can be broken down into broad categories on the basis of demand.

Table 2.1: Mineral content in Higher Plant dry matter

Average mineral content in the dry matter of Green Plant	
Element	Content g/kg
O	440
C	420
H	60
N	30
K	20
P	4
Other elements together	26

Macronutrients

There are 6 macronutrients which are all required in significant quantities by plants, i.e. 2-30 g/kg of plant dry matter.

Table 2.2: Concentration of Major nutrients in Leaf tissue

Macro nutrient concentration (%) in Mature leaf tissue

Nutrient - Primary	Deficient	Sufficient	Excess (toxic)
Nitrogen	< 2.50	2.50 – 4.50	> 6.0
Phosphorus	< 0.15	0.20 – 0.75	> 1.0
Potassium	< 1.00	1.50 – 5.50	> 6.0
Nutrient - Secondary			
Sulfur	< 0.20	0.25 – 1.00	> 3.0
Calcium	< 0.50	1.00 – 4.00	> 5.0
Magnesium	< 0.20	0.25 – 1.00	> 1.5

Table 2.3: Major nutrients are applied as fertilizers and crops take up them as mineral radicals;

Nitrogen [N]	Taken up as NO_3^- or NH_4^+
Phosphorous [P]	Taken up as $H_2PO_4^-$ and HPO_4^{2-}
Potassium	Taken up as K^+
[b] Secondary nutrient	Applied for certain crops some soils
Sulphur [S]	Taken up as SO_4^{2-}
Calcium [Ca]	Taken up as Ca^{2+}
Magnesium [Mg]	Taken up as Mg^{2+}

Micronutrients

There are 7 micronutrients which are just as important as macronutrients to the health of the crop but which are required only in very small amounts i.e. 0.3-50 mg/kg of plant dry matter.

Table 2.4: Micro nutrient concentration in higher plants

	Micronutrient concentrations in leaf tissue (ppm) of Higher plants		
	Deficient	Sufficient	Excess
Boron	5 – 30	10 – 200	50 – 200
Chloride	<100	100 – 500	500 – 1000
Copper	2 – 5	5 – 30	20 – 100
Iron	<50	100 – 500	>500
Manganese	15 – 25	20 – 300	300 – 500
Molybdenum	0.30 – 0.15	0.1 – 2.0	>10.0
Zinc	10 – 20	27 – 100	100 – 400

Table 2.5: Forms (chemical radicals) of nutrients as absorbed by plants

C	CO_2 from the atmosphere and HCO_3 – from soil solution
H	Comes with other oxidized elements, water and as O_2
O	Comes with water and other elements
N	Ionic forms from soil solution & gas NH_3
P	Phosphates from soil solution
K, Mg, Ca, Mn, Cl, Na	Ionic form from soil solution
Fe, Cu, Zn, Mo	Chelated or ionic form from soil solution
B, Si	Boric acid or borate, Silic Acid

In addition to these essential mineral elements, there are some nutrients which are beneficial to some plants.

Table 2.6: Other essential elements required for plant growth

Sodium[Na]	Taken up as Na+ [partly replaces K^+ on some crops]
Silicon [Si]	Taken up as silicate etc. for strengthening cereal stems
Cobalt [Co]	Mainly useful for "N" fixation of legumes
Aluminium [Al]	May be beneficial to some crops.

Yield Limiting Factors

Under traditional farming systems using only organic manures, crop yield is most often restricted by the availability of Nitrogen. Therefore yield can be increased by simply adding N. Once N ceases to be the limiting factor however, yield will be restricted at a slightly higher level by the next deficient nutrient, usually P or K or any one of the listed above micro-elements.

For maximum yields to be achieved, the optimum level of nutrient needs to be available to the plant in advance of actual growth. However, too much nutrient, or "Luxury Supply", which cannot be utilized, is a waste. Excessive fertilizer application can also reduce crop growth due to toxicity.

The role of the 3 major nutrients

i) Nitrogen

Essential for the production of proteins in the plant. Proteins are one of the building blocks of plant tissue and so N is very important for vegetative growth. N is an integral component of chlorophyll, which is responsible for photosynthesis and gives the plant its green colour.

Deficiency of N therefore shows a yellowing of the green part of the plant, usually older leaves first, as N is relocated within the plant from old tissue to new tissue. Nitrogen is absorbed by the plants in the ionic forms – NO_3 (Nitrate) and NH_4^+ (Ammonium).

Table 2.7: Rating of Soils with their Nitrate-N contents.

Soil Test rating	Nitrate Nitrogen (ppm)
Very low	0 to 3
Low	4 to 10
Medium	11 to 20
High	21 to 40
Very High	> 41

ii) Phosphorus

Essential for growth and elongation of the plant, both above and below ground, P is particularly important in the early stages of growth for the development of a healthy root system on which the later growth of the plant will depend.

Phosphorous is also essential for seed and fruit development, and maturation.

Deficiency of 'P' shows a blue/red colouration on the leaves, and stunted growth. The ripening of fruits is also affected.

Low mobility and solubility in soil. It is absorbed by the plants in two forms; H_2PO_4 -- and HPO_4 .

At low soil pH P is associated with Fe, Mn and Al metals and at high pH with Ca.

Table 2.8: Rating of soil by their P content (Olsen method)@.

Soil Test Rating	P soil test at pH 7 (ppm)	P to be added (kg/ha)
Very low	0 to 6	60-80
Low	7 to 15	40-60
Medium	16 to 26	30-60
High	27 to 45	10-25
Very high	46 plus	0 or Maintenance

Source: Ankerman and Large (1980)

iii) Potassium

Essential in the regulation of transpiration, also plays a role in the plants resistance to disease, and is also important for the development and quality of fruits and seeds.

Deficiency of K shows first as chlorosis and scorching of leaf tips and margins. Small, light and cracked fruit may also result. K is highly mobile in plants and is regularly re-transported from older tissues to new growth.

Table 2.9: *Soil test rating for K, Mg and Ca given as % exchangeable cations vs CEC.

Soil test Rating	K (% CEC)	Mg (% CEC)	Ca (% CEC)
Very low	< 0.75	< 0.85	< 10
Low	0.75-1.5	0.85- 2	10-20
Medium	1.5- 3	2-5	20-50
High	3-5	5-10	50-80
Very High	> 5	> 10	> 80

* Soils Bulletin (1980)

Table 2.10: Generalized Symptoms of Plant Nutrient Deficiency and Excess

Plant Nutrient	Type	Common Visual symptoms
Nitrogen	Deficiency	Light green to yellow appearance of leaves, especially older leaves; stunted growth; poor fruit development.
	Excess	Dark green foliage which may be susceptible to lodging, drought, disease and insect invasion. Fruit and seed crops may fail to yield.
Phosphorus	Deficiency	Leaves may develop purple coloration; stunted plant growth and delay in plant development.
	Excess	Excess phosphorus may cause micronutrient deficiencies, especially iron or zinc.
Potassium	Deficiency	Older leaves turn yellow initially around margins and die; irregular fruit development.
	Excess	Excess potassium may cause deficiencies in magnesium and possibly calcium.
Calcium	Deficiency	Reduced growth or death of growing tips; blossom-end rot of tomato; poor fruit development and appearance.
	Excess	Excess calcium may cause deficiency in either magnesium or potassium
Magnesium	Deficiency	Initial yellowing of older leaves between leaf veins spreading to younger leaves; poor fruit development and production
	Excess	High concentration tolerated in plant; however, imbalance with calcium and potassium may reduce growth.
Sulfur	Deficiency	Initial yellowing of young leaves spreading to whole plant; similar symptoms to nitrogen deficiency but occurs on new growth.
	Excess	Excess of sulfur may cause premature dropping of leaves.
Iron	Deficiency	Initial distinct yellow or white areas between veins of young leaves leading to spots of dead leaf tissue.
	Excess	Possible bronzing of leaves with tiny brown spots.

3

Soil Fertility

Although the application of fertilizer to even fairly fertile soils can give a great increase in crop yield, soil fertility is not simply a matter of nutrient content. It is dependent on a number of complex of issues:

Soil depth

The volume of soil from which the crop draws its nutrition varies according to the rooting depth of the crop. However hard or rocky layers can restrict natural root development; this is particularly the case with micro irrigation systems.

Soil structure

Particle size and soil aggregate size and distribution, effect the size and distribution of pores and therefore the availability of air and water to the roots.

Soil pH

This is an indicator and regulator of chemical and biological processes in the soil.

pH Changes element's solubility

e.g

pH>7 = Phosphorus and micro elements fixation occurs.

pH<7= Increase in micro element solubility

pH very acidic Al and Mn toxicity and Ca/Mg deficiency.

Soil Electrical conductivity (EC)

An index to determine the concentration of ions in the soil solution

Normally presented in: Deci-semen's/Meter (dS/m) or millisiemens/cm (mmhos/cm)

Normal irrigation water ranges at 0.2-1.2 mmhos/cm

Adding soluble salts and ions to water increases EC

EC is highly affected by Cl^- and NO^{3-}

a) **Nutrient Status :** The content and availability of nutrients in the soil for crop growth.

b) **Storage Capacity :** For available and soluble nutrients from both natural sources and fertilizer application.

c) **Humus Content:** The release of nutrients from organic matter in the soil is dependent on the quality of the material as well as favorable soil conditions for its break-down.

d) **Microbial Activity :** Soil organisms act as agents in the break-down of organic matter and nutrient release into the soil.

e) **Toxicity :** Substances in the soil harmful to crop growth can reduce the yield regardless of nutrient availability.

Soil Factors Affecting Nutrient Availability

As can be seen, many factors can affect the availability of nutrients to the plants. Soil, as a natural living medium for plant growth does not conform to simple uniform rules and so there are no rules of thumb which can determine exact nutrient availability over a growing season and therefore how much fertilizer needs to be applied to a particular crop. Some of the major factors which affect soil fertility are discussed here.

A) Cation Exchange Capacity [CEC]

Very small particles of clay and humus in the soil carry a negative electrical charge, which attracts and holds positively charged cations. Cations include some of the nutrients essential for plant growth such as:

a) Potassium $[K^+]$

b) Magnesium $([Mg^{2+}]$

c) Calcium $[Ca^{2+}]$

d) Manganese $[Mn^{2+}]$

e) Ferrous $[Fe_2^+$ and $Fe^{3+}]$

Copper $[Cu^{2+}]$ and Zinc $[Zn^{2+}]$

as well as H^+, Na^+ and Al^{3+}.

Cations in the soil solution and absorbed on root surfaces can be exchanged with those attached to clay or humus particles in a process known as cation exchange.

The total quantity of cations that can be exchanged by both clay and humus in a soil is referred to as the cation exchange capacity [CEC]. Ion exchange is not significant on soil particles larger than silt and so the CEC of a sandy soil with low organic matter [O.M.] would be very low. Once this concept is understood, it can be seen that this process of exchange will have a direct influence on nutrient availability to the plant. Divalent cations, e.g. $Ca2+$ are more strongly held than monovalent cations, e.g. $K+$. Exchange of cations takes place on an equivalent weight basis.

The CEC varies according to soil type, as already mentioned low O.M. soils with particles predominantly greater than silt size will not have a significant CEC, but within the group of clay and high O.M. soils with high CEC, there is considerable variation according to the type of clay and the form of organic matter.

Some typical CEC values for several soil types are shown in table below.

Table 3.1: Typical CEC values for different soil types.

Soil Type	CEC [Milliequivalent/100 g]
Sand	<4
Sandy loam	4-8
Loam	10-18
Silt loam	10-25
Clay and clay loam	20-35
Organic soils	50-200

The CEC is usually expressed in milli equivalent per 100 grams [1 milli equivalent = equivalent weight/1000].

B) Base Saturation

A number of cations are called base cations, these are cations which help to reduce soil acidity, and include K^+, Ca^{2+}, Mg^{2+} and Na^+ .

The percentage of the CEC which is occupied by these base cations, rather than acidifying cations

such as H^+ and Al^{3+} is referred to as % base saturation. This is important because of its effect on nutrient retention in the soil and also on soil pH, and therefore nutrient availability.

C) Soil pH

The term pH is used to indicate the level of H^+ Ions in the soil solution, and is therefore a measure of actual acidity rather than potential acidity, [the potential acidity would be related to the number of H+ Ions attached to the soil particles also]. The pH scale ranges from 0 to 14 and is defined as the log of the reciprocal of H+ Ion concentration in the soil solution. 1

$pH = -Log\ (H^+)$

On the pH scale, 7 is neutral. This is the point at which H+ is equal to OH-, below 7, H^+ Ions are greater in number than OH^- Ions and as H^+ is an acidifying ion, the soil is acidic. Over pH 7 the reverse is the case and the soil is alkaline.

As the scale is logarithmic, every unit increase or decrease represents a 10 fold increase or decrease in H+ ion concentration.

The pH of the soil is important because of its effect on a number of chemical and biological processes in the soil. The width of colour cylinder corresponds to the quantum of availability.

The direct effect of pH on a number of major and secondary nutrients has been well established. Phosphate availability in particular is closely associated with pH, maximum availability of P occurring at pH 6-7 with availability rapidly declining either side of this range.

In acid soils reaction of P with soluble aluminum, iron or manganese reduces P availability, while in alkaline soils it is the formation of insoluble forms of calcium phosphate which limits phosphate availability.

The availability of a number of micronutrients, increases at pH levels below 7.0, and for this reason micronutrient deficiencies, particularly of Fe, Mn and Zn are commonly seen on alkaline soils.

Molybdenum, however, becomes insoluble if the pH falls below 6.0, and toxicity problems may arise below pH 5.0 as high levels of aluminum and manganese are solubilized. Soil pH also indirectly affects the supply of nitrogen to the crop.

In general anions, including nitrate and phosphate, are taken up at a higher rate under weak acid conditions, but under moderate to strongly acid conditions the microbiological processes which convert ammonium N into nitrate N and which fix gaseous N in the soil are suppressed. This results in a drop in the availability of nitrate N which can affect plant growth.

Other effects of pH are to influence the occurrence of certain plant diseases, and also weed control and crop injury from residual herbicides.

The optimum pH range varies from crop to crop. Most crops prefer a pH range between 5.5 and 7.5, but a few do well in either acid or alkaline soils.

Table 3.2: Optimum pH values for various crops

I. Cereals	pH	II. Legumes	pH
Maize	6.0-7.5	Beans	5.5-7.5
Millet	5.5-6.5	Ground nut	5.5-6.5
Rice	4.0-6.0	Soyabean	5.5-7.0
Sorghum	6.0-7.5		
Wheat/Barley	6.0-7.5		
III. Other crops			
Sugarcane	6.0 - 8.5	Tomato	6-6.5
Cotton	6.0-8.0	Peppers	6-6.8
Tea	4.0-6.0	Banana	4.0-8.5
Safflower	6.0-8.0	Grape	6.0
Citrus	5.0-6.0	Papaya	6.0-6.5
Cashew	4.5-6.5	Coconut	6.0-7.0
Mango	5.0 - 6.0		

What Causes Acidity in Soils?

There are a large number of factors which may contribute to the soil solution being largely saturated by H^+ Ions rather than the basic cations.

Soil is formed by the weathering of a parent rock material, and so if the parent material contains very few basic cations then it is natural that the soil will be acidic. Even though the soil originally contained sufficient basic cations, they have been lost from the soil by processes such as leaching, soil erosion and removal in harvested crops over long periods of cultivation. The soil pH close to plant roots can also be reduced by the release of H^+ as NH_4^- is taken up.

In addition, carbonic acid can be formed by carbon dioxide exhaled from plant roots, combining with water in the soil. Leaching is a particular problem on sandy, free draining soils which have a very low CEC, for this reason sandy soils are more prone to acidity than clay soils.

Acidity can be caused also by the use of acid or acid forming fertilizers, the use of such fertilizers should therefore be avoided in acid soils, but can have obvious benefits in alkaline soils.

Raising the pH of Acid Soils

The pH of acidic soils is usually raised by "liming" the soil, that is adding one of a number of materials which contain cations such as Ca_2^+, Mg_2^+ or K^+ together with basic anions such as CO_3^-.

One of the most common materials used is calcium carbonate [$CaCO_3$]. The "buffer" capacity of the soil, [the soils ability to maintain a stable pH], can also be increased by the addition of organic matter.

What Causes Alkaline Soils?

There are 4 recognized causes of alkalinity and the methods for dealing with each type is different. It is important therefore to recognize the nature of the alkalinity before it can be dealt with.

High Lime Soils

Soils rich in Ca_2^+ and Mg_2^+ can result naturally in areas of very low rainfall where very little leaching of cations occurs, or if the parent rock is rich in these cations, e.g. chalk, limestone etc. In countries where liming is practiced on slightly acid soils, over application can also result in alkalinity problems.

In high lime soils with a pH < 7.5, the soil can be neutralized by applying acidic or acid forming fertilizers such as urea or ammonium forms of N. Sulphur or sulphur compounds can also be used.

If the high pH of a soil is a result of the presence of coarse particles of $CaCo_3$, then the use of acidifying fertilizers will not achieve a long term reduction in the soil pH, as more and more calcium will simply be exposed and reacted to release Ca_2^+ into the soil solution. However, the use of acid fertilizers in these soils can create lower pH conditions in the crop root zone on a temporary basis, and so increase the availability of essential nutrients to the plant.

a) Saline Soils

Alkaline soils which contain excess salts [E.C. > 4 mmhos/cm] are often found in arid or semi-arid areas, particularly where irrigation has been carried out over a number of years.

Such soils, which have a pH ranging from 7->8.5, are deteriorated by the gradual buildup of salts in the root zone. This is caused by the deposition of salts in the soil by the irrigation water applied. This is a particular problem where the irrigation water is of low quality, i.e. contains high levels of salts.

Provided that there are no hard layers preventing drainage these soils can be improved by leaching out the excess salt by over irrigation with water low in Na^+. Where natural drainage is not possible, the installation of artificial drains can be effective. Otherwise in situations where the soils are too thin for under drains or the water table is high, then planting crops on raised beds and/or irrigation with drip systems may reduce the problem.

b) Sodic Soils

Large amounts of sodium [Na^+] in the soil can ruin the soil structure, impeding drainage and thus making the soils much more difficult to reclaim. The installation of drainage systems together with the application of Ca_2^+ as gypsum [$CaSO_4 \cdot 2H_2O$] to improve soil structure can improve the chances of reclamation by making possible the leaching of the excess Na+.

c) Saline Sodic Soils

These are soils which have both a high salt content [E.C. > 4 mmhos/cm] and excess Na^+. These soils, which will have a pH < 8.5, can be easier to reclaim than simple sodic soils because they tend to retain better soil structure.

Factors Affecting Root Development

The supply of nutrients in the soil is meaningless unless the plant can take them in through the root system. The development of the roots is therefore essential for good crop growth, and they require the right conditions in which to develop.

a) Soil Structure

Soil structure refers to both the basic soil type, (particle size and type) and also to the arrangement of these particles into "aggregates". Aggregates are groups of soil particles which have stuck together to form crumbs of soil and their formation is linked to the level of clay and organic matter in the soil.

Aggregates are important in determining the formation of macro-pores in the soil. These pores play an important role in the supply of air and water to the root system, the infiltration of water into the soil, available water retention within the soil and drainage of excess water out of the root zone.

b) Soil Depth

Different crops have different patterns of root development which determine rooting depth. This also depends on an adequate soil depth and the absence of hard layers in the soil, such as plough pans which may impede growth.

c) Irrigation

In arid regions, the only supply of water for much of the year may be from the irrigation system. Root development is therefore dependent on the correct distribution of water and the depth of infiltration.

Uneven irrigation due to poorly designed or maintained systems can have a dramatic effect on root penetration. This is particularly the case with micro irrigation systems where only a part of the soil is being wetted.

d) Organic Matter

Organic matter has long been in use as a fertilizer, but it has many more useful effects than simply the release of nutrients.

O.M. acts as a soil conditioner. On heavy clay soils it acts to create a more open soil structure with better aeration and a higher percentage of available water, while on a sandy soil it acts to bind the large soil particles into aggregates and to increase water retention. In addition O.M. can increase the CEC of a soil thus providing a buffer against pH fluctuations.

e) Aeration of Soil

Poor aeration has a retarding effect on root development. Root growth and the metabolism of moisture absorption are retarded under conditions of low oxygen tensions.

4

Fertilizers

Organic Fertilizers and Manures

Farmers have been using organic waste as fertilizers for a very long time. As previously mentioned, the application of O.M. to a soil can have number of beneficial effects on soil structure, C.E.C. and microbial activity. However, organic manures mainly originate from the waste products of animal and plant production and although rich in carbon compounds, they are usually relatively poor in available plant nutrients.

The nutrient content of organic manures varies widely depending on their source and moisture content. Farm yard manure (F.Y.M.) is one of the most common organic fertilizers used. Others are concentrated animal manures, and crop residues such as straw.

Table 4.1: NPK content in various organic manures#

S. No.	Name of Manure	% of Nutrient		
		N	P_2O_5	K_2O
1.	FYM (well decomposed)	0.8	0.65	0.5
2.	Goat/Sheep Excreta	0.6	0.5	0.7
3.	Compost	0.5	0.65	0.88
4.	Night soil	1.3	1.1	0.35
5.	Groundnut cake	7.10	1.40	1.30
6.	Cotton seed cake	6.40	2.80	2.50
7.	Castor cake	4.5	1.7	0.7
8.	Neemcake	5.0	1.0	1.50
9.	Bone meal	3.50	21.5	-
10.	Fish meal	4.10	0.90	0.30

#One Hectare crop of Green manure fixes 60-90 kg Nitrogen/ha per year in soil.

While the quantity of N, P and K in the manure can be measured by simple analysis, the actual availability of nutrients to the crop is much more difficult to establish. Only a fraction of the total N in F.Y.M. is available as urea or mineral N, and the remainder as organic compounds which can remain in the soil for a very long time.

P is also only released very slowly though K tends to occur as soluble K+ and therefore can be quickly utilized. The application of low quality organic manures can actually reduce the amount of N available to the crop, as microorganisms active in the breakdown of the waste utilize available nitrate N for the breakdown process.

Important criteria for the value of organic fertilizers are

1) Dry matter content > 60 to 65%.

2) Total and easily mineralizable humus.

3) Organic matter content > 60%.

4) Total and quick acting N content.

5) Carbon: Nitrogen ratio < 20.

6) Content of toxins harmful to crop growth.

While organic fertilizers have an important role to play in crop production, high levels of production are unlikely to be achieved by the use of manuring alone. In addition there are a number of organic products in the market which are of doubtful value to the crop.

Mineral Fertilizers

Mineral fertilizers have been developed in order to supplement the natural supply of nutrients in the soil and thus make it possible to achieve higher yields. These fertilizers are produced either in single nutrient, multinutrient or complex forms, and contain nutrients as mineral salts or in organic forms which are easily mineralized.

a) Single Nutrient, (Straight) Fertilizers

Containing predominantly one major nutrient, although often a secondary nutrient may be present, e.g. Ammonium Sulphate.

b) Multinutrient Fertilizer

A mix of more than one single nutrient fertilizer to produce a substance containing at least 2 and possibly all 3 of the major nutrients.

c) Complex Fertilizer

Where 2 or more of the nutrients are chemically combined e.g. Urea Phosphate.

Fertilizers containing more than one major nutrient are often referred to as NPK fertilizers and the nutrient content is expressed as % weight in the order of N, P_2O_5 and K_2O e.g. 19:19:19 indicates that the fertilizer contains 19% N, 19% P_2O_5 and 19% K_2O by weight. [The usual practice is to quote the oxide forms of P and K, but sometimes the elemental form is used, it is well worth checking exactly what the figures quoted refer to].

Mineral fertilizers [except specialized slow release products], contain Nitrogen in the form of urea, ammonium or nitrate. Phosphorous fertilizer is in the form of phosphate, and Potassium as either potassium chloride or potassium sulphate.

Secondary and micronutrients can also be applied when needed as mineral fertilizers, either in conjunction with one or more of the major nutrients or individually.

It should be stressed that whether nutrients originate from organic or mineral fertilizers, they are taken up in the same form by the crop. The advantage of applying mineral fertilizers is therefore that higher rates can be applied and greater control can be exercised over plant nutrition.

d) Water Soluble Fertilizers and Liquid Fertilizers

The availability of nutrients to the crop from mineral fertilizer depends on the solubility of the nutrient forms used in that fertilizer. Some fertilizer products are more soluble than others under soil conditions, and so the rate of release of the nutrients applied into the soil solution can vary.

With traditional mineral fertilizers, large doses of fertilizer are applied to a crop between 1 to 5 times in a year. Usually all the P and K are applied at planting with part of the N, and additional N is subsequently applied in several smaller doses through the growing cycle. The fertilizers applied in this way will be gradually released into the soil solution and be taken up by the crop.

A percentage of the fertilizer applied will be wasted either through leaching, volatilization, or fixation in the soil. The exact processes depend on a range of factors such as soil type and pH, and climate conditions such as temperature and rainfall, but utilization of NPK fertilizers applied as solids to the soil tends to be in the range of:

N : 50-60%,   P: 10-30%,   K: 50-60%.

The technique of fertigation, which is the application of fertilizers already in solution through the irrigation water, offers great potential for improving these efficiencies.

By applying nutrients in an available form at regular intervals through the irrigation water, preliminary research in Indian Universities indicates potential for saving 25-30% of fertilizer through greater efficiency of use.

i) Low Chloride Fertilizers

Some crops are sensitive to high amounts of chlorides in the soil, which can have the effect of depressing yield, effecting quality, and causing direct injury to the crop.

Such crops are termed chlorophobic, and it is therefore of benefit to use fertilizers which contain very low levels of chlorides (often termed "chloride free" fertilizers]. The cheapest form of potassium fertilizer is muriate of potash [potassium chloride], and in many cases, this fertilizer has been widely used on irrigated land in arid areas, leading to a buildup of chlorides in the soil. For such areas, or where chlorophobic crops are to be grown, the more expensive potassium nitrate (KNO_3) or potassium sulphate (K_2SO_4) based fertilizers should be used.

Most field crops are not chloride sensitive and a few crops such as oil palm and coconut actually benefit from higher levels of chlorides.

Table 4.2: A list of chlorophobic crops

Grapes
Strawberry
Pineapple
Fruit trees (particularly Citrus)
Cucumber and some other vegetables
Onion
Floriculture Crops, e.g. Roses
Cotton
Tobacco

Calculating fertilizer requirements

Every farmer asks "how much fertilizer should I apply to this crop". There is no single answer as the issues surrounding plant nutrition are many and varied.

Every crop is different and has different requirements through its growing cycle. Every field is different, because of soil type, past cultivation and fertilization practices.

Local knowledge and experience of soil conditions is valuable in achieving at least medium yields by concentrating on limiting nutrients, but achieving and maintaining high yields and more economic use of fertilizers requires a more systematic approach. Soils differ widely in their ability to provide nutrients for plant growth. Nutrient availability in a soil depends not only on the amount of nutrient reserves, but also the degree of mobilization and fixation of nutrients in the root zone.

Because of the variable factors involved, it is necessary to assess empirically the nutrient status of soils and plants in order to provide guidelines for profitable fertilizer use. There are a number of approaches which we can use to do this.

a) Fertilizer recommendation based on yield goals

The primary purpose of using fertilizers is to increase production and quality of product from a crop. Although a low yield farmer may wish to become a high yield farmer overnight, it is unlikely that he will be able to achieve his aim as high yields are dependent on a wide range of management factors. Fertilizer alone will not achieve this aim and a thorough knowledge of all production factors is needed.

Targeted increases in yield should therefore be kept within reasonable limits on a year by year basis. It is reasonable to expect a low yield farmer to achieve medium yields within one or two seasons of changing his farming system, and similarly for a medium yield farmer to move towards higher production, but not to achieve a miracle overnight!

Having established a reasonable yield target, it is a simple matter to calculate the approximate nutrient requirement of a crop to achieve that target. By utilizing the research data which has been generated to quantify the nutrient uptake and the removal per unit yield of various crops, the nutrient requirements can easily be worked out.

It is much more difficult however, to determine how much fertilizer needs to be applied to the soil to meet these requirements. The amount of fertilizer required to achieve the target yield will almost always be greater or smaller than the crops requirements for the following reasons.

1) The loss of nutrient from the soil as a result of leaching or vitalization.

2) The fixation of nutrients in the soil by chemical and physical processes.

3) Gains from the mineralization of organic matter in the soil, and release of "fixed" nutrients into the soil solution.

4) The nutrient status of the irrigation water.

5) The fixation of atmospheric nitrogen by soil bacteria.

b) Experimental data collection

The best way to produce accurate fertilizer recommendations is to conduct scientific fertilizer trials on the main soil types in a climatic region.

This is the only way to take into account all the local conditions which may affect fertilizer uptake and nutrient availability. But scientific trials are an expensive and time consuming procedure beyond the capabilities of almost all farmers. They can however be conducted by Government agencies, universities and some commercial companies. Fertilizer trials generally take a number of years to complete and even then the data is only specific for the trial site.

c) Soil Tests

Analysis of nutrient levels in the soil, in combination with the targeted yield goal can give a much more accurate picture of fertilizer requirements.

Soil analysis is by no means perfect. Based on the concept of available nutrients, it only gives a snap shot of the nutrient status of the soil at the time of sampling. It can however, indicate other parameters affecting soil fertility such as organic matter content, soil pH, and salinity, thus enabling the fertilizer recommendations to take those factors into account.

d) Interpretation of Soil Analysis

Once the soil samples have been tested, a report will be produced indicating the fertility status of the soil. For this information to be useful it is important to be able to make sense of it and adapt fertilizer recommendations as a result.

For most situations only N, P and K application will be required. The table below indicates the likely effect of nutrient levels in a soil on crop yield if no fertilizer is applied, (for P, K & Mg).

Soil analysis gives nutrient levels in the soil as a concentration e.g. ppm or mg/kg.

Table 4.3: Interpretation of Soil test data for Macronutrients

Supply class	Expected relative yield without fertilizer %	Available (Extractable) nutrients (mg/Kg Soil)		
		P	K	Mg
Very low	50	<5	<50	<20
Low	50-80	5-9	50-100	20-40
Medium	80-100	10-17	100-175	;40-80
High	100	18-25	175-300	80-100
Very High	100	>25	>300	> 180

Usually in India, soil analysis results will be quoted in Kg/ha rather than mg/Kg. This is done by calculating the weight of soil in one hectare which the soil analysis represents, and then multiplying the concentration X weight of soil = weight of nutrient. Therefore for a standard 15 cm soil sample the above figures translate below (Table 4.4) as Kg/ha.

Table 4.4: Interpretation of soil test for macronutrients (Kg/ha) units.

For 15 cm soil sample: (at bulk density 1.2 = 1.8 million kg soil/ha)	Supply Class	P	K
	Very Low	<9	<90
	Low	9-16.0	90-180
	Medium	16.0-30	180-315
	high	30-45	315-540
	Very high	>45	>540

e) Adapting Fertilizer Recommendations to Soil Conditions

Where fertilizer application rates for a particular crop in a particular area have been established, either through research trials, experience, or by calculating nutrient requirement from the target yield, they can be adjusted according to the soil analysis report in the following way.

Table 4.5: Fertilizer requirement for achieving targeted yield.

Soil Supply Class fertilizer	Expected relative yield without	Fertilizer adjustment required
Very low	50%	Increase by 50% or more
Low	50-80%	Increase by 25%
Medium	80-100%	Apply recommended dose only
High	100%	Decrease by 25%
Very High	100%	Decrease by 50% or more

Table 4.6: An Example : Banana fertilizer recommendations from research trials.

N (Kg/ha)	P_2O_5 (Kg/ha)	K_2O (Kg/ha)
440	178	440

If available Potash in the soil = 350 Kg per hectare (K_2O = High levels in soil) therefore decrease K_2O application by 25%

440 Kg/ha less 25% = 330 kg

N	P_2O_5	K_2O
Fertilizer application Kg/ha		
440	178	330

This is an accepted and useful technique for amending fertilizer applications according to soil analysis reports.

However if we look at the fertilizer uptake of Banana it is in the region of:

N (Kg/ha)	P_2O_5 (Kg/ha)	K_2O (Kg/ha)
450	89	1350

So it can be seen that the N application is close to the university recommendations whereas the P_2O_5 and K_2O application are very different.

The reasons for this are that in most Maharashtra soils:

1. Phosphate availability is low due to the high pH.

2. Potash is often naturally available in large quantities and so the fertilizer is only supplementary to the natural supply.

i) Nitrogen

In contrast to Phosphate and Potash, the level of available 'N' in a soil changes very quickly. Available N is in the forms of NO3- (Nitrate) and NH4+ (Ammonium), Nitrate N is extremely mobile in the soil and can be lost by leaching if irrigation efficiency is poor. It can be useful therefore, to test soils for available N regularly in order to amend fertilizer applications and avoid either over supply or shortages.

Because of the mobility of available N, a larger part of the crops total requirement will need to be applied through fertilizer. N which is available in organic forms, (other than Urea), in the soil is only slowly released into the soil solution in an available form.

Available N is not the total soil N.

Total N = Organic N + Available N. and as organic N is not readily available to the plants, be careful to check that the available 'N' figure given, really does mean available.

Testing For Available N

Experience in the U.S.A suggests that soil samples for testing available N should be taken from above and below plough depth for reliable results. This means that between 30 and 60 cm soil depth is tested, depending on the rooting depth of the crop.

Much of the available 'N' in Indian conditions will be released during the monsoon, when also a significant amount will be lost by leaching or loss to the atmosphere, (denitrification). For this reason it may be of value to test for available N at the end of the monsoon. Available N shown at that time can be considered as 'Fertilizer already applied', when calculating future applications.

For example

Table 4.7: Steps in estimation on N

Banana crop requires	440 kg/ha 'N'
Fertilizer applied during monsoon	100 kg/ha 'N'
Post Monsoon application	340 kg/ha 'N'
Post monsoon soil analysis shows	40 kg/ha 'N' – available
Amended Fertilizer application	300 kg/ha N.

ii) Organic Matter and Nitrogen

It is quite common for nitrogen requirements to be calculated on the basis of soil organic matter alone. This is quite a crude calculation full of assumptions which may introduce an error into the result, however it can be useful as a guide, and made more reliable by taking into account the presence of available N in the soil. The nitrogen release from soil organic matter can therefore be calculated as follows: e.g. for a soil bulk density 1.3 kg/liter

Table 4.8: Estimation of N release from Organic matter

Hence a furrow slice 15 cm deep x 10,000 m^2	1,500 m^3 or 1,500,000 liter
1,500,000 x bulk density of 1.3 kg/ltr	1,950,000 kg top soil per hectare
1,950,000 kg x % organic matter x % "N" in O.M x soil release factor	"N" released.

1) The % organic matter can be easily measured in laboratory tests.

2) The % N in organic matter is usually taken as 10%.

3) The soil release factor varies according to soil type and temperature and represents the fraction of the nitrogen content released from O.M. per year. For U.S. conditions, it has been calculated as:

Clay : 0.02, Loam : 0.025, Sand: 0.03.

If a loam soil has an organic matter content of 1% the nitrogen you could expect to become available in a year would be: 19,50,000 Kg X 0.01 X 0.1 X 0.025 = 48.75 kg/hectare N.

Of this 48.75 kg N/ha however, not all would be utilizable by the crop and the timing of availability also cannot be predicted.

In addition, the Carbon: Nitrogen ratio of the organic matter has a marked effect on the release of available N. Organic matter with high C:N ratios, such as straw and bagasse, actually tend to temporarily reduce nitrogen availability in the soil. It may be suspected therefore that a simple measure of the organic carbon content of the soil is not a reliable basis for making fertilizer recommendations.

The actual release of N which can be and will be utilized by a crop cannot be easily predicted and can best be established by field experiments. It would be safe to assume that only around 60% of "N" released will actually be utilized by the crop.

Using Target Yields

Data from experiments has established nutrient uptake in kg/tonne yield for most of the major crops. Nutrient requirement for the crop for the target yield can therefore be calculated, e.g. Sugarcane

Table 4.9: Estimation of nutrient requirement

	N	P_2O_5	K_2O
Nutrient uptake kg/t yield (Zende 1983, India)	1.2	0.46	1.44
So, for a 120 t/ha yield			
Nutrient needed =	144	55	173 kg/ha
But for a 150 t/ha yield			
Nutrient needed =	180	69	216 kg/ha

These figures tell us what the crop needs, but the amount of fertilizer to be applied will depend on the level of nutrients available in the soil and the expected fertilizer use efficiency.

Sugarcane - e.g. 150 t/ha target yield	N	P	K
Required Nutrient, as elements	180	30	180
Soil analysis	40	8	125
(Available 'N' 60% usable)	(24)		
Balance required	156	22	55
Fertilizer use efficiency	60%	30%	50%
Element applied as fertilizer	260	73	110
	N	P_2O_5	K_2O
As oxides	(260)	167	132

N.B. These calculations are made using expected fertilizer use efficiencies for conventional fertilizers.

Maintaining Soil Fertility

This will supply sufficient nutrient to give the target yield, but the release of nutrients from the soil may not be able to compensate for the nutrient removal in the first year. A soil test in the following year may show lower reserves of nutrients in the soil.

Eventually, the P & K levels applied will have to match the uptake and removal of the nutrient in the harvested yield.

It is better therefore, to maintain soil fertility at a medium level so that there is some buffer capacity in the soil to prevent periods of plant hunger.

Where soil nutrient status is high, application rates can be reduced, but where soil nutrient status is low, additional fertilizer can be applied over a number of years to improve soil fertility. Where nutrient status is medium, applying the full amount will compensate for removal.

If we recalculate our 150 t/ha Sugar cane example in this way:

Table 4.10: Components in nutrient estimation.

	N	P	K
Required nutrient (as element)	180	30	180
Soil analysis (60% of 'N' usable)	40	8	125
Medium analysis levels	—	16	180
Extra nutrient required	(24)	8	55
Total nutrient required	156	38	225
		(27% increase)	(25% increase)
	N	P_2O_5	K_2O
As Oxides	156	87	270
Fertilizer use efficiency	60%	30%	50%
Fertilizer to be applied	260	290	540 kg/ha

Normally this extra application would not be made in one year due to the expense, but over a number of years.

N.B. These calculations are made using expected fertilizer use efficiencies for conventional fertilizers.

Amending fertilizer recommendations to take into account of organic fertilizers added to the soil.

The recommendations for fertilizer added to the soil for crop growth can be reduced to account for the contribution made by organic manures etc. Manures can vary widely in their total "N" content and ability to release that "N" into the soil in an available form. If the manure can be analyses, then a more certain prediction of nutrient contribution can be made, but in the absence of analysis an estimate can be made based on general figures:

Table 4.11: Mineral contents of common organic manure

	N%	P_2O_5 %	K_2O%
FYM	0.5	0.65	0.5
Compost	0.5	0.65	0.88
Night Soil	1.3	1.1	0.35
Neem cake	5.0	1	1.5

It should be noted that about 30% N, 50-60% P_2O_5 and 80-90% K_2O will be available to the crop in the first year.

Materials such as straw which are reincorporated into the soil have a high Carbon: Nitrogen ratio and are therefore likely to tie up soil available nitrogen, (at least initially), in the microbial decomposition of the waste material. As a general rule this decomposition of dry plant material will require about 1 kg extra N per 100 kg of waste material.

The fixation of N by leguminous plants can be in the region of 50-280 kg per hectare depending on the crop soil conditions and effectiveness of rhizobial, [N fixing bacteria], activity.

Secondary Nutrients

The need to apply secondary nutrients is less usual than for the major nutrients, but demand is growing as yields increase. The need is dependent on soil type, crop requirements and the level of targeted yield.

a) Calcium

The main requirement for calcium application is usually related to raising the pH of acid soils, where plant deficiencies occur, soil application of calcium can often be only partially effective. Foliar sprays are often the easiest and most effective means of applying calcium in response to deficiency.

The ratio between calcium and magnesium in the soil is of some importance in order to avoid Mg deficiency, the Ca : Mg ratio should not be greater than about 10:1.

b) Magnesium

Magnesium deficiency is becoming more common under intensive cropping systems as NPK fertilizers increase yields and therefore put a greater demand on the Mg supply.

Mg_2^+ concentration is often higher in the soil solution than for example K^+ but it appears that plant uptake of Mg_2^+ is generally less efficient than for K^+. So magnesium deficiencies are associated not only with a shortage of Mg_2^+ in the soil solution but also with high levels of K^+. This may be due to competition for uptake by the roots. Tests have indicated that low K+ levels lead to high Mg_2^+ uptake while high K^+ has the opposite effect.

Magnesium deficiency is very distinctive and if observed can be rectified either by soil applications or foliar sprays, however, if "K" levels in the soil are high, foliar application may be the best option.

c) Sulphur

The need for sulphur containing fertilizers has also increased as average crop yields have risen. Considerable amounts of sulphur can be released from the breakdown of organic matter however, and so the greatest likelihood of deficiency is in low OM soils.

Micronutrients

Micronutrients are essential nutrients which are generally required in very low concentrations by the plant. The need for application is generally restricted to certain crops grown in difficult soils, but as yields increase the need is growing. The effect of deficiency on yield and crop quality can be quite serious.

As the crop requirement is generally small, foliar sprays of sulphate or chelated form of the nutrients can be very effective. Chelates can be applied informs suitable to the soil conditions and thus remain available for crop growth [EDTA chelate is suitable for acid soils, EDDHA chelate is suitable for alkaline soils].

Plant Testing

Analysis of samples of tissue from the growing crop and of fertilizer applications during the growing period. Nutrient concentration in the plant tissue is generally a good indicator of nutrient status in the soil. This is an important indicator for good plant nutrition although it can be expensive and requires great care in the collection and handling of samples.

Interpretation of the samples is based on experimental data for each individual crop, which has identified "critical values" for macro or micronutrient content in the plant tissue, outside of which range an unacceptable reduction in yield is likely to occur.

Results are usually expressed in %, grams per kg or ppm dry matter. For this technique to be useful in amending fertilizer applications during the growing season, it is essential that testing laboratories are able to give the result within a few days of sampling.

Other Methods of Diagnosis

a) Optical - Plant Observation

Nutrient deficiency in plants, if serious enough, will show physical signs in the colour or vigour of the crop.

With enough experience it is possible to identify the deficient nutrient from the symptoms which the plant is showing.

Basically a healthy crop with adequate nutrition shows what is known as "full green colour". Any deviation from that colour may be caused by a lack of nutrients. Once the cause has been identified the nutrient application can be amended. However, if the deficiency is very minor, it may not show clear visible symptoms and yet an effect on yield. This is termed" hidden hunger". In addition, if more than one deficiency occurs at the same time, the symptoms shown by the plants may be difficult to interpret.

In either of these cases, either suspected deficiency without obvious symptoms, One can take practical steps to determine the cause by applying the likely deficient nutrients to a few plants, as foliar sprays for quick results, and observing the effects over a period of several days. If an improvement is seen then the problem has been correctly identified and the same treatment can be given to the whole crop.

b) Nutrient Balance Sheets

Once soil fertility has been established by means of soil tests, then recording crop yield against fertilizer application can enable the farmer to retain an understanding of his soils fertility status from season to season. Such a procedure, while greatly simplified in terms of the processes affecting soil fertility, has been successfully used in some countries as a practical tool to guide fertilizer application between periodic soil tests.

The conditions for which this example has been calculated are foreign to India, nevertheless there is no reason why the concept should not be adaptable to suit local soils and climatic conditions here.

c) Value: Cost Ratio

It is essential that for fertilizers to be worth using, they must bring some economic gain to the farmer.

As a guideline, the financial value of the extra yield of crop obtained: cost of the fertilizer used, should be at least 2:1 to make good economic sense.

Queries before planning for fertilizer application

Before a farmer should opt for the application of fertilizer to increase his income, he should be sure that there are no other limiting factors on his yield which will prevent him gaining the benefits he expects from fertilizer usage. He should ask himself:

1) Are all other agronomic practices satisfactory, eg. irrigation management

2) Are the basic requirements for soil fertility in place? eg. suitable pH, good soil structure.

3) Is enough information available on soil fertility status, crop nutrient requirements and optimum timing of fertilizer application to enable the farmer to make the right decisions about quantity and timing of application?

4) Are the availability of fertilizer suitable for the crop and the soil. Is the supply of fertilizer reliable, and is the technology used to apply the fertilizer understood by the farmer?

Unless the answer to all these questions is yes, then the expected benefit from the use of fertilizer may not be achieved.

5

Fertigation

Fertigation is the technique of supplying dissolved fertilizer to crops through an irrigation system. When combined with an efficient irrigation system (like drip method of irrigation) both nutrients (type and quantity) and water (volume or rate) can be manipulated and managed to obtain the maximum possible yield of marketable product from a given quantity of these inputs.

Compared to traditional surface irrigation systems, fertigation targets a small volume of soil; 20-30% of the total volume of soil. With drip irrigation the wetted area is bulb shaped and called "wetted bulb" and it is here the water and fertilizer are targeted. The size and shape of the wetted bulb depends on soil texture and the type of micro irrigation system used. Because water and minerals are concentrated in the wetted bulb active root growth also takes place here. The combination of supplier and demand concentrating in the soil in wetted bulb enforces the precision control of crop growth. This is the crux of the drip/micro irrigation technology. The anchor roots of the crop grow out of the wetted bulb and they are functionally different.

The purpose of Agriculture is to maximize yield and quality of crops and minimize the costs of production while maintaining sustainability. Two most important prerequisites to achieve all of these are the optimal and balanced supply of water and nutrients. Tuning of plant nutrient supply with uptake by crops is essentially an art and is made doable by fertigation technology.

Often, solid fertilizer side-dressings are timed to suit management constraints rather than the horticultural requirements of the crop. For example, most farmers will have experienced the dilemma of spreading fertilizer the day before heavy rain and then wondering how much of the fertilizer is either washed from the crop in run-off or leached below the root zone.

Continuous applications of soluble nutrients in very small doses overcome these problems, save labour, reduce compaction in the field, result in the fertilizer being placed around the plant roots (rhizosphere) uniformly and allow for rapid uptake of nutrients by the plant. To capitalize on these benefits, particular care should be taken in selecting fertilizers and injection equipment as well as in the management and maintenance of the irrigation system.

With drip irrigation, only part of the soil volume is wet and contributes fertilizer ions to the plant. However, only about 40% of the soil volume actually supply ions to the plants. Therefore, fertigation becomes the only way for obtaining high fertilizer efficiency when micro irrigation is adopted.

Brief history of fertigation technology

Origin of fertigation can be attributed to hydroponics, a method of growing crops in soil-less culture. Hydroponics is an ancient technology used as early as in the Hanging Gardens of Babylon and the Floating Gardens of Aztecs of Central America. In both these cases fresh water rich in oxygen and nutrients were pumped into the hydroponic systems.

Coming to modern scientific agriculture era, in early 19th century a Frenchman, Jean Baptist Boussingault who identified 9 elements required for plant growth when he supplied these elements in liquid form to plants grown in an inert medium. He identified not only the elements but the rates and combinations required for optimum plant growth. By 1938, English scientists Hoagland and Arnon through their independent research established various nutrient formula for plant growth. From 1925 the glass house growers turned to both hydroponics (growing in liquid medium) and soil less culture (Peat, Sand, Sawdust or Vermiculite media) wherein both these cultures required nutrients in liquid form to be applied. So fertigation was borne.

It is also interesting to note that the first large hydroponics system was established by the US army during 2nd World war for feeding the men in the war front!

By 1950 mixing fertilizers with irrigation water was used in limited scale in flood, surface and furrow irrigations. Mostly N fertilizers, gaseous ammonia, ammonium nitrate were mixed with irrigation water. Because the efficiency of irrigation was low so the fertilizer use efficiency too. With the introduction of surge irrigation fertilizers were injected into the surge valves. This has improved the efficiency to some extent. Electrical pumps and fertilizer mixing tanks were then introduced for the first time in Netherlands when they went into glass house crop production.

By 1960 fertigation found a new and efficient partner; micro irrigation. Because of the small volume of wetted soil in drip irrigation, an adequate supply of nutrients to the rhizosphere required the synchronization of water and nutrient placement.

Fertigation technology also saw the evolution of injecting systems. Initially fertilizer tanks were used to introduce fertilizer solution into the irrigation flow. Later more uniform injections were attained by the use of venturi and suction pumps. Nowadays, computer controlled injectors achieving very high precision application and pH, EC sensors and control devices are available.

The introduction of liquid fertilizers in 1970 boosted the idea of fertigation in micro irrigation to higher levels. Matching with this development, is the refinement of control systems to stop reverse flow of fertilizer solution to water source since most water sources are dual purpose sources- drinking and irrigation.

Equipment for Fertigation

Though fertigation as a technology is possible in all types of irrigation methods, as mentioned above, in this manual we are describing the method only for pressurized irrigation methods like drip (drippers, and other micro- emitters like micro-sprinklers, jets and sprayers) and sprinkler systems. In pressurized irrigation there is by definition, pressure within the irrigation network. Injecting fertilizer solution into such systems requires generating a pressure differential to overcome the internal pressure. How can we achieve this without hindering the irrigation flow is the crux of the technology for fertigation equipment. Several types of equipment are developed over the years with this objective in mind.

Fertilizers for Fertigation

There is a large range of chemical fertilizers offered for fertigation. Nonetheless, the suitability of a fertilizer for fertigation depends on several of its properties; especially its solubility in water.

Solid fertilizers completely soluble in water at field ambient temperature are suitable for fertigation. Most conventional fertilizers that are soluble and several specialized fertilizers in water soluble form are available at present for fertigation.

Characteristics of fertigation friendly fertilizers

Soluble in water at ambient temperature.

Only completely water soluble fertilizers can be used for fertigation. The fertilizer solution should be prepared at ambient temperature using the irrigation water. Attempt to heat the solution to increase solubility will result in precipitation of salt when mixed with the general flow of irrigation water in the system.

Similarly, a fertilizer that is highly soluble in summer season may not dissolve enough in winter.

Solutions should not interact with metal used in the containers or FT. All containers for mixing and transferring solutions should be plastic or non -metallic.

i) Solubility of fertilizer

Table 5.1: Solubility of certain fertilizers at different temperatures.

Fertilizer	Chemical Formula	Solubility g/liter - Temperature °C			
		0	10	20	30
Urea	$CO(NH_2)_2$	680	850	1060	1330
Ammonium Nitrate	NH_4NO_3	1183	1580	1950	2420
Ammonium Sulphate	$(NH_4)_2SO_4$	706	730	750	780
Calcium Nitrate	$Ca(NO_3)_2$	1020	1240	1294	1620
Potassium Nitrate	KNO_3	130	210	320	460
Potassium Sulphate	K_2SO_4	70	90	110	130
Potassium Chloride	KCl	280	310	340	370
Mono Potassium Phosphate	KH_2PO_4	1328	1488	1600	1790
Di-ammonium phosphate	$(NH_4)_2HPO_4$	429	638	692	748
Mono Ammonium Phosphate	$NH_4H_2PO_4$	227	295	374	464
Magnesium Sulphate	$MgSO_4$	260	308	356	405

ii) Compatibility of fertilizer

Fertilizers should be compatible with irrigation water. Water may contain divalent cations: Calcium or Magnesium; and if phosphate fertilizers are dissolved they form precipitates.

While mixing more than one fertilizer the fertilizers should be chemically compatible. For example Calcium containing fertilizers are not compatible with most other fertilizers (Table 5.2). They are to be fertilized alone without mixing with any other chemical.

Table 5.2: Chemical compatibility of different fertilizers

Urea	Ammonium Nitrate	Ammonium phosphate	Calcium nitrate	Potassium Nitrate	Potassium chloride	Potassium sulphate
Ammonium Nitrate						
Ammonium sulphate						
Calcium Nitrate		xx				
Potassium Nitrate						
Potassium Chloride						
Potassium sulphate		x	xx		x	
Ammonium phosphate			xx		x	
Iron, zinc, copper, manganese sulphate			xx		x	
Iron, zinc, copper, Manganese chelate			x			
Magnesium sulphate			xx		x	
Phosphoric Acid			xx		x	
Sulfuric acid			xx		x	
Nitric acid						

Urea	Ammonium phosphate	Iron, zinc, copper, manganese sulphate	Iron, zinc, copper, mangane sechelate	Magnesium sulphate	Phosphoric acid	Sulfuric acid	Nitric acid
Ammonium Nitrate							
Ammonium sulphate							
Calcium Nitrate							
Potassium Nitrate							
Potassium Chloride							
Potassium sulphate							
Ammonium phosphate							
Iron, zinc, copper, manganese sulphate	Xx						
Iron, zinc, copper, Manganese chelate	X						
Magnesium sulphate	Xx						
Phosphoric Acid			x				
Sulfuric acid							
Nitric acid			xx				

X Reduced solubility; XX – incompatible.

iii) Interaction between water and fertilizer in fertigation

Most irrigation water has intrinsic salt content and a certain osmotic pressure because of it. The osmotic pressure of the irrigation water would increase when fertilizers are dissolved. Therefore the total osmotic pressure of the solution should be kept in mind while fertigating. Because a very high osmotic pressure in the rhizosphere will be counterproductive for plant growth and yield. At a raised osmotic pressure, plants use more energy for water and nutrient uptake and this extra energy is expended at the cost of yield. However, one would not regularly measure the osmotic pressure of the solution. Instead the EC (Electrical Conductivity) of the solution is measured.

(OP is related to EC; OP= 0.36 x EC)

The acidity or alkalinity of fertigation solution is expressed as pH. This indicates the corrosion hazard of the solution. pH below 6.0 increases corrosion and pH above 7.5 causes salt precipitation. Nevertheless, solutions with lower pH would concurrently clean the drippers of any clogging salt.

EC of the solution indicates the ionic strength (IS). IS = 0.013 x EC. Therefore the EC, pH and nutrient concentration of fertilizer solutions are estimated. These factors are very critical in precision agriculture and can lead to crop damages if not adequately controlled. These are of immense importance in green house fertigation.

Table 5.3: EC, pH and Nutrient concentrations of fertilizer solutions $

Fertilizer	Chemical Formula	Nutrients	Conc. mg/l	EC ds/m	pH
Ammonium Nitrate	NH_4NO_3	N	280	0.7	2.00
Calcium Nitrate	$Ca(NO_3)2$	N	280	2	5.5
Ammonium Sulphate	$(NH_4)2SO_4$	N	280	1.4	4.5
Urea	$CO(NH_2)2$	N	280	2.7	7
Mono-ammonium Phosphate	$NH_4H_2PO_4$	N	140	0.4	4.7
		P	310		
Phosphoric Acid	H_3PO_4	P	310	0.4	2.3
Mono-Potassium Phosphate	KH_2PO_4	P	310	0.7	4.6
		K	390		
Potassium Chloride	KCl	K	390	0.7	7
Potassium sulphate	K_2SO_4	K	780	0.2	7
Potassium nitrate	KNO_3	N	140	0.7	7
		K	390		
Magnesium Sulphate	$MgSO_4$	Mg	240	2.2	6.9

$ It is recommended to carry out a jar test to assess the suitability of the fertilizer for fertigation.

iv) Fertilizer "jar test"

This will help you avoid problems due to incompatibility or checks solubility of fertilizer.

Whether you begin with water-soluble or liquid fertilizers, dissolved chemicals such as phosphates, calcium, and magnesium can react together or with the irrigation water. This can lead to insoluble chemical combinations precipitating in the water. These precipitates can clog the emitters. There are also fertilizer compatibility charts available, such as the one below. But they may not list the fertilizers you are considering to use.

The "jar test" is easy. The key is to approximate the dilution rate that you expect to be injecting through the drip system.

One only need the following to carry out a jar test

The injection rate (lph)

The drip system delivery rate (lph)

The stock fertilizer or fertilizer combinations that you will be using (stock solution to water applied ratio)

A jar with a sealing lid

The water that you use for irrigating (use the buffer (6.5pH) irrigation water)

Characteristics of water soluble fertilizers

Water soluble fertilizers come in two main forms- liquids and solids. Water soluble solids are easier to transport and comes in very convenient packages. Liquid fertilizers are increasingly becoming extinct because of the difficulty in transporting large volumes of liquids. Water soluble solids should possess the following characteristics for their suitability for fertigation.

- Completely water Soluble
- Acidic in Nature
- Corrects Soil pH / maintain soil pH
- Nutrients should be in readily available form
- Should not contain Na & Cl
- Can be applied in smaller quantity
- Application should not provide chances of fixation of P & K
- Should not cause soil deterioration
- Should improve plant up- take of nutrients
- Suitable for foliar feeding

Interaction of fertilizer compounds with irrigation water.

Because irrigation water carries soluble salts at various concentrations. The EC and pH of these irrigation water would change the EC and pH of the fertilizer solution leading to factors of specificity. For example, high concentrations of ammonium sulphate first acidify the water. The SO4 anion may combine with Ca2+ dissolved in water and precipitates CaSO4. This would cause clogging. While fertilizers like Urea and Ammonium nitrate do not tend to interact with salts dissolved in irrigation water and hence fertigation of these fertilizers would not pose any danger.

Knowledge of these interactions on a fertilizer to fertilizer basis is essential in preparing the fertigation program for crops and different irrigation water sources.

Interaction of fertilizer with soil and release of nutrients.

Plant nutrients applied through fertigation move with soil and react with soil (other growing media in case of soil less culture). The form of fertilizer applied in fertigation has a marked effect on its distribution in the wetted bulb.

i) Nitrogen

The form of nitrogen containing radical (NO_3 or NH_4) will decide the location, infiltration rate and spread of the nutrient movement in the wetted bulb.

NO_3- Nitrogen is very mobile and flows down with the water.

Ammoniacal nitrogen is slow and remains in the top soil layer.

Because of this trait, excess drip irrigation after fertigation event would leach out NO_3 nitrogen beyond the root-zone. It is also advisable to use smaller doses on nitrogen and increase the frequency of fertigation.

In case of ammonium source of nitrogen, special care should be taken not to accumulate excess ammonium in the top soil layers by adjusting the dose and time (frequency) of application. Excess ammonium is toxic to many crops. Ammonium in the top layer of soil can also be lost through volatilization. These are also likely to be fixed in clayey soils. If the pH of soil solution is alkaline then more of the ammonia (NH_3) is lost by volatilization. As a solution to these issues it is recommended to use only up to 20% of the N as ammonia source. In hydroponics and soil-less culture (green house cultivation) use ammonium for only up to 10% of the total N requirement.

Urea is highly soluble in water. It will move with the water in the soil till it is hydrolyzed by the enzyme urease to ammonium carbonate; $CO(NH^{)2} + 2H_2O = (NH_4)^2CO_3$.

Ammonium carbonate is unstable and decomposes into ammonia and carbon dioxide;

$$(NH_4)^2CO_3 = 2NH_3 + CO_2 + H_2O.$$

The ammonia (NH_3) may be adsorbed on soil surfaces Or dissolve in water as the ammonium cation , NH_4^+ which in turn may be adsorbed by the cation exchange sites of the soil.

Likewise, if ammonium salts are applied instead of Urea, then also the NH_4^+ ions are formed and adsorbed into the cation exchange sites.

NH_4^+ will be oxidized to NO_3 by microbes in the soil. The rate of nitrification depends on environmental conditions. Autotropic bacteria converts NH_4^+ into nitrite (NO_2) first and then to Nitrate (NO_3). This whole process is known as nitrification.

$$2NH_4^+ + 3O_2 = 4H_+ + 2H_2O + 2NO_2^-$$

$$2NO_2^+ + O_2 = 2NO_3^-$$

As we can see oxygen is required for these reactions and Hydrogen ions (H+) is released; thereby acidifying the soil environment.

Soil texture affects the rate of nitrification because of the fact that different textures contain different amounts of water. At a soil water potential of 0.01 bar (water saturated soil) with no air in the soil, nitrification stops because bacteria assisting in nitrification need oxygen. Similarly, nitrification also stops at soil water potential of 15 bar (very dry soil), because of lack of available water for the bacterial activity. This clearly shows the essentiality of optimum soil water status for maximum fertilizer efficiency.

Soil pH also influences nitrification. The optimum pH range for nitrification is pH 6.6 and 8.4.

Nitrate (NO_3^+) will move with irrigation water through the soil (growth media) because it does not react with soil components. In a fertigated system, movement of water and thus of NO_3^+ can be sufficiently controlled so that leaching below the root active zone can be minimized. This results in very low Nitrate leaching into the ground water.

However, in high organic soils and when oxygen is limiting, Nitrate gets reduced to first N_2O (nitrous oxide) and then to gaseous Nitrogen. This process, denitrification is also assisted by microbes and utilizes glucose as an energy source.

ii) Phosphorus

Water soluble sources of P for fertigation are Ammonium phosphate, Potassium phosphate, and phosphoric acid.

However, in the soil these salts may react with di- or tri-valent cations and form less soluble compounds. MKP (monopotassium phosphate) was found to be an effective source for P fertigation (and of course K).

Polyphosphate sources after coming in contact with soil/media are hydrolyzed by enzymes and gets into very complicated chemical reactions. For example ammonium Polyphosphate solution contains tri-poly phosphate, orthophosphate and pyrophosphate and other higher polymers. The end product of these reactions are orthophosphate.

$H_5P_3O_{10} + 2H^2O = 3H^3PO^4$.

The temperature of the soil/medium, water content, and pH will affect the rate of this hydrolysis. Because Phosphate is highly immobile it remains in the top layer of the soil immediately below the dripper whatever may be the soil type. It is therefore advised to fertigate it in smaller doses throughout the P requiring stages of the crop.

iii) Potassium

Potassium fertilizers used in fertigation are readily soluble and the K remains as a positively charged ion in non reactive growth media and sandy soils.

In clayey soils, however most of the K is retained as exchangeable, non-exchangeable and fixed K. Exchangeable K is readily available to plants. As plant roots take up K from the soil solution it is replenished first by the exchangeable K and this in turn is replenished by the fixed K.

Chloride source of K should not be used in saline soils. Sulphate or nitrate of K should be more appropriate in all cases.

iv) Calcium, Magnesium and Sulphur

All these three elements are highly mobile and their concentration tends to increase at the boundaries of the wetting bulb. Excess irrigation would result in their leaching away from the root zone.

Micronutrients (Boron, Copper, Iron, Molybdenum, Zinc and Manganese)

The pH of the soil affects the availability of these micro elements. Chelates of Copper, Iron, Manganese and Zinc give best performance in fertigation. The anionic micronutrients, boron and molybdenum are generally applied as borate and molybdate salts.

e) Nutrient distribution in soil by fertigation

Vertical distribution of the applied nutrients in soil depends on two factors;

1) Rate of movement of irrigation water in the soil profile

2) Properties of the soil.

The nutrient dissolved in water moves down into the soil profile with the water. Only if the ions enter into chemical reaction with the soil/or growth media the movement is obstructed. For example, phosphates may get precipitated in the presence of Ca, Fe, or Al in the soil solution. K, Mg, or NH_4 ions of the nutrient solution may be retained by the Cation Exchange (CE) sites of the soil. In such cases the nutrient ions cannot move freely with water.

Nitrate ions, however do not get obstructed in any of these chemical processes.

Soil texture influences the rate of movement of nutrients in water. In soils with high coarse particles or soilless growth medium containing coarse particles the movement of water and the dissolved ions is even and unobstructed. In fine textured soils; clay, silt combinations, water movement will squarely depend on the size of the pores in the soil. In such soils, a dissolved nutrient ion like nitrate will form a gradual concentration gradient from the point of application to the last depth of its arrival with water.

The pressurized water supply in drip irrigation linked to controlled fertilizer injection may insure a more even spatial and temporal water and fertilizer distribution. Thus the volume of field soil wetted by drip fertigation and enriched with nutrients varies according to;

Quantity of water applied, and Water holding capacity of the soil.

f) Fertigation through Sprinkler Irrigation

Sprinkler irrigation is suitable and economical for certain crops- low value cereals, pulses and leaf and tuber crops. In this case diverse types of nozzles (water emitters) are available to choose from to match the water application rate to the rate of infiltration into the soil.

Uniformity of water application by sprinklers is essential for using their system for fertigation. Utilization of Solid-sets of sprinklers or self-propelled systems (Central pivot irrigation machine, for example) minimizes the role of labour and associated variability in the distribution of water and fertilizer.

All types of sprinkler systems can be used for fertigation; but with the following caution

Avoid of corrosion of metallic components in the sprinkler system by contact with corrosive fertilizers.

Take care of the plant canopy from scorching by caustic fertilizers spread through the sprinklers.

Use plastic bodied sprinklers as much as possible to prevent damage to the system by the fertilizers.

Fertigation requires to be 100 % covering the entire crop. Sprinkler laterals should be placed in such a way to ensure the total coverage of fertilizer placement.

If sprinklers are chosen for orchards, use always under-canopy sprinklers (micro sprinklers/jets etc.) and ensure that each tree gets the same volume of water. This would result in uniform fertilizer application.

Always operate the sprinklers at the design pressure; a loss of pressure at sprinkler point would

a) reduce the throw distance of water

b) result in incomplete wetted area cover.

Both of these would reduce the efficiency and uniformity of fertigation.

What is Fertilizer Tank?

"It is a small pressure vessel designed for application of soluble fertilizer/ chemicals into the irrigation line as per the crop requirement by creating a pressure difference in the drip system".

Technical Features

It is made of high grade mild steel & heavy duty engineering plastic.

Separate valves are provided on the inlet & outlet to control the injection rate.

Effective mixing of fertilizers / chemicals is due to innovatively designed inlet pipe, extended upto the base of the tank with elbow mechanism.

Suitable size of manhole of 8" width makes easy operations i.e. pouring of fertilizer stock solution & cleaning of internal surface of tank after each fertigation cycle.

The fertigation application can be done at a minimum pressure difference of 0.2 kg/cm^2 and tank can sustain maximum working pressure of 10 kg/cm^2 (ML) & 3 kg/cm^2 (PL).

Fertilizer tank storage capacities are 30, 60, 90, 120 & 160 litres in metal, 30, 60 & 90 litres in plastic model.

On demand fertigation tank can be supplied in stainless steel material with any capacity.

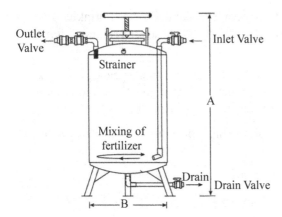

Fig. 5.1: Schematic diagram of fertigation through a Fertilizer tank (FT)

Installation

Fertigation tank can be installed in many ways but general practice followed are as follows:

Installation by means of fertigation manifold on mainline or separate submain.

Installation by means of Sand Filter Manifold.

Note: Care should be taken while installing the fertigation tank that there should not be any leakage, loose / defective connections & it should be placed on level platform.

Operation & Maintenance

Operation in the fertilizer tank takes place by creating differential pressure in the pipe line by throttling or partial closing of valve, thus by diverting the flow of mainline into the tank containing a stock solution, hence the water and dissolved fertilizer both with certain level of concentration are fed in the downstream. Hence one should apply the fertilizer doses by referring the following performance chart.

Table 5.4: Performance Chart of Fertilizer tank

Inlet / Outlet Pressure Difference	Minimum operation time in minutes				
kg/cm²	30 Liter	60 L	90 L	120 L	160 L
0.2	12	15	25	30	40
0.3	09	10	15	15	20
0.5	07	08	10	10	15

Note: Tested under standard test conditions.
Source: Courtesy Jain Irrigation Systems Ltd.

- Operation time is given as guidelines only and actual timing can be changed according to the density of stock solution and other field conditions.

Advantages

- Very simple to operate, the stock solution does have not to be pre-mixed.
- Easy to install and requires very little maintenance.
- Easy to change fertilizers.
- Ideal for dry formulations.
- No electricity or fuel is needed.

Disadvantages

- Requires pressure loss in main irrigation line or a booster pump.
- Proportional fertigation is not possible.
- Limited capacity.

What is Venturi Injector?

"It operates by creating differential pressure in the pipe line by means of throttling or partial closing of valve; hence the fluid under high pressure is converted into a high-velocity jet at the throat of the convergent-divergent nozzle which creates a low pressure at that point. The low pressure draws the suction fluid into the convergent-divergent nozzle where it mixes with the motive fluid & it is injected in the pipeline"

Technical Features

It is made of heavy duty engineering plastic which as excellent chemical resistance.

Innovatively designed convergent, throat and divergent performs accurate & hassle free operation of fertilizer injection.

Separate valve is provided at suction side to prevent the draining of water during non-operating condition.

To avoid the entry of solid or foreign particles in the drip system, strainer is provided at the suction side.

Venturi models are available in different sizes i.e. ¾", 1", 1 ¼", 1 ½" & 2".

It is highly efficient, effective & compact differential pressure injection device.

It is one of the economical and low cost fertigation options available today.

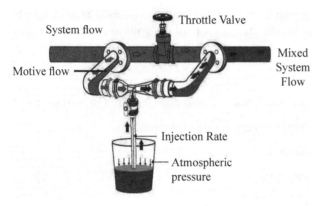

Fig. 5.2: Schematic diagram of working of a venturi

Installation: A Ventury can be installed in many ways but general practices followed are as follows:

Installation by means of fertigation manifold on mainline or separate submain.

Installation by means of Sand Filter Manifold

Note: Care should be taken while installing the Venturi injector, that, there should not be any leakage, loose / defective connections which in turns will effect on injection properties. Venturi requires exact amount of motive flow and pressure difference for accurate fertigation operation

Advantages

- It is very simple & compact device with no moving parts.
- It can be easily installed and maintained over the period.
- It can be controlled with different injection rates as per our requirement.
- Injection rates can be controlled with the help of metering valves place on the injection line.
- It is suitable for proportional as well as quantitative fertigation application.

Disadvantages

- Requires pressure loss in main irrigation line or a booster pump
- Quantitative fertigation is difficult
- Automation is difficult.

Table 5.5: Injection rates for Venturi Injector ¾"

Pressure Inlet	Outlet	Motive Flow thru venturi injector	Injection Rate
kg/cm²	kg/cm²	Liter per minute	Liter per hour
	0	8.5	77
	0.2	8.4	70.8
1	0.4	8.3	47.7
	0.6	8.2	23.5
	0.8	8.2	-
	0	10.5	73.8
	0.2	10.4	64.2
	0.4	10.4	68.1
1.5	0.6	10.3	50.5
	0.8	10.3	30.1
	1	10.3	-
	0	12.4	70.8
	0.2	12.3	70.2
	0.4	12.2	70
	0.6	12.2	64.8
2	0.8	12.2	49.8
	1	12.2	37.3
	1.2	12.2	30.5
	1.4	11.9	-
	0	14	69.3
	0.2	13.9	68.4
	0.4	13.9	67.7
	0.6	13.8	66.1
	0.8	13.8	60.6
2.5	1	13.8	54
	1.2	13.8	47.1
	1.4	13.8	30.9
	1.6	13.7	23.6
	1.8	13.7	-

Note: Test conducted under standard test conditions with venturi injector installed on bypass.
Source: Product catalogue Jain Irrigation

What is Injector Pump?

"Injector pump is a device which has a piston mechanism to inject the fertilizer solution into the drip system, such type of injector pumps are also called as proportional fertigation pumps."

Technical Features

Fig. 5.3: Injector pump for fertigation

Applications: medication for livestock, cleaning or disinfecting fluid lines, sanitizing water systems, car wash, cost effective fertilizing, and pest control for crops and plants.

Features

- Innovative compact design which operated without electricity
- To avoid entry of solid particles in system, additional suction filter is provided
- For accurate injection, special adjusting lock arrangement provided
- Made of strong plastic material with high chemical resistance.
- Choice of selection on the basis of desired proportions,
- Adjustable from 0.4% to 4%.

Available in 3 general models to provide operational flexibility,

1) Manual (ON-OFF)

2) Hydraulic (ON-OFF)

3) Electrical (ON-OFF)

Maximum operating pressure 5 kg/cm^2 (70 psi). High pressure model (6 kg/cm^2) are also available.

Table 5.6: Available in different sizes and capacity

Size	Flow Rate (m³/hr)	Device setting	Motive Flow (lph)	Injection rate (lph)
3/4"	2.5	0.4 - 4.0%	20-2500	0.08 - 100
3/4"	3.5	1.0 - 10.0%	50-3500	0.5 - 350
1.0"	5.0	0.5 - 5.0 %	200-5000	1.0 -250
1.5"	10.0	1.0 - 5.0 %	500-10000	5.0 - 500
2.0"	25.0	1.0 - 5.5 %	2000-25000	20.0 - 1375

Source: Courtesy- Jain Irrigation systems Ltd.

Advantages

- Very accurate, for proportional fertigation
- Less pressure loss in the line
- Easily adapted for automation
- Works at low flow rate.
- No electricity needed
- Easy to adjust the dosing rate.

Disadvantages

- Relatively expensive
- Complicated design, including a number of moving parts, so wear and breakdown are more likely.

Advanced Fertigation Machine- JAIN NUTRICARE

Fertigation plays an important role to achieve the high yield and productive results. Jain irrigation system ltd. has developed "NutriCare" an advance fertigation machine for precise fertigation. It works as bypass nutrients injection machine. By pass injection machine can be connected with existing irrigation system.

It is designed for dosing of nutrients in quantity and time based on continuous and pulse operation. Proper suction and mixing of nutrients in irrigation water results the uniformity in crop growth, stem size, fruit growth and fruit quality. Unique hardware differentiates it from other dosing devices. Low power consumption and user friendly operation is the simplicity of machine.

NutriCare is the only device which maintains the EC and PH of irrigation water. This device is equipped with EC & PH monitoring and transmitting controller. Multiple nutrients selection in adequate quantity and delivery as per crop requirement and its predefined injection pattern makes it superior than any other bypass injection machine.

Most of the parts in NutriCare are non-moving part which makes it maintenance free or low maintenance. Components of the machine are selected having suitability and durability in injection of fertilizer and acid for long life. NutriCare is environment friendly noise free and integrated machine which can be installed at any place.

Components of NutriCare

1) Pressure Booster pump

2) Fert pump

3) Fert pump manifold

4) Non return Valve

5) Solenoid valve

6) Rotameter (flow meter)

7) Needle Valve

8) EC-PH monitor

9) Sampling Cup

10) Electrode (EC & PH)

11) Pressure Gauge

12) Pressure sustaining valve

13) Pressure switch

14) Electric Control panel

15) Body Frame

16) Double action Air release valve

17) Hose pipe

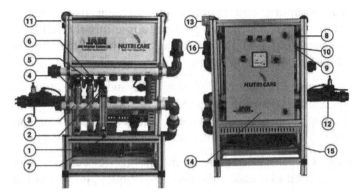

Fig. 5.4: Fertigation machine- Jain NutriCare -Front and back view.
(*Courtesy*- Jain Irrigation Systems Ltd.)

Features & Specifications

1) **Pressure booster pump:** Range from 1.5HP to 5HP- 3-phase pump, selected in adequate size to match the motive flow and Creates the pressure difference between inlet and outlet of fert pump..

2) **Fert pump:** Capacity from 160 LPH to 1000LPH. Fert pump acts on pressure difference. It is designed with high graded PVC material to make it free from corrosion and degradation by imposing acid.

3) **Fert pump Manifold:** Available in 2 to 7 set of connection ports. Made of UV stabilized PVC material.

4) **Non return Valve:** Provided with each fert pump set to prevent the reverse flow.

5) **Solenoid valve:** Direct Acting 24VAC operated. This valve is being used to execute the fertigation process. Solenoid valve is operated through external PLC or controller. It is designed and tested for frequently operation without any fail.

6) **Rotameter (flow meter):** Flow measuring range 50 to 1000 LPH. to monitor the suction flow of fert-pump.

7) **Needle valve:** Arrangement with all flow meter to regulate the flow of fertilizer. Suction of the fert-pump can be adjusted manually as per change of demand of fertilizer in crop.

8) **EC-PH monitor:** Plays an important role to monitor the EC and PH of delivered final solution. Receives the signal from Electrodes and transmit to irrigation controller in the form of analogue signal 4-20mA. Monitor provides the facility of calibration by using push button in simple steps.

9) **Sampling cup:** Sampling cup is made of corrosion free material and collects sample from the Main pipe line to multiply the fertilizer proportionally. Sampling cup design makes it special for precise fertigation and has frequent sampling capability.

10) **EC & PH Electrode:** PH electrode in Glass tube and EC electrode in PVC high graded material are designed to bear the pressure up to 10bar. Leak proof sealing arrangement provides accuracy in measurement. Connectors and shielded cable provides the accuracy in measuring data.

11) **Pressure Gauge:** Glycerin filled pressure monitoring device. Assembly with two way selection to measure the pressure of inlet pipe and outlet pipe.

12) **Pressure Sustaining valve:** Hydraulically operated pressure sustaining valve with adjustable pilot. Sustaining valve is adjusted in higher pressure than pipe line pressure. Booster pump create the pressure to open and close the sustaining valve at pre-set pressure rating.

13) **Pressure switch:** Pressure switch provided with NO/NC dry contact. Pressure switch protect the booster pump from dry run operation. Starts and dry run stop of the pressure booster pump takes place on the basis of inlet line pressure. If the pressure in the main line will be lower than preset pressure value than pressure switch will not allow to start the pump.

14) **Electrical control panel:** Electrical control panel provides the safety and phase reversal protection to pump. Short circuit protection with MCB breaker assembly. Water proof control panel with Earthing port, cable entry port appropriate gland and robust enclosure. Control panel is equipped with pressure booster pump starter.

15) **Body frame:** High graded corrosion resistant aluminum frame. Light weight, strengthened, fitting slots and spacious body frame. Modular design capability, assembling and dismantling can be easily done on site.

16) **Double Action air release valve:** Size ½" to 2". It protects the system from injury due to unwanted air from main line. Incoming air affects in the suction of fert-pump.

Working of NutriCare

NutriCare Smart fertigation device is controlled by Jain SPIRIT PRO irrigation controller. Controller has Digital output, and Analog input which controls the Booster pump, fertilizer pump and solenoid valve. Controller is responsible for making irrigation and fertigation schedules.

NutriCare machine is selected considering total flow of the system, Pressure in main line, maximum requirement of fertilizer per fertilizer pump and operational time of NutriCare etc.

Controller sends the operational sequence to NutriCare machine as per following methods:

1) Time based continuous operation

2) Volume based proportional operation

3) EC –PH based operation

NutriCare receives the signal from controller and starts operating direct acting solenoid valve. Controller also sends the command to booster pump and the booster pump checks the status of pressure switch. Booster pump starts working when the pressure of main line reached the set value of pressure switch.

In time based continuous operation, fertilizer injection process takes place up to the pre-defined time/fertigation schedule.

In volume based operation, water meter and nutrients counter meters are the part of operational method. Injection process is based on main line real time flow and quantity of fertilizer required. It injects the fertilizer in pulses.

In EC-PH based fertigation process, this fertigation method is adopted to achieve the set value of EC & PH. This method is independent of fertilizer quantity and time. Variable quantity is injected but finally machine maintains the required EC & PH.

In Above all the process flow of the fertilizer can be adjusted manually by using needle valve. Flow meter provides the real time flow of the fert- pump. Pressure booster pump creates the pressure difference in inlet and outlet of fert-pump. Fert-pump starts suction from fertilizer tanks. Booster pump takes irrigation water from main line and after adding of fertilizers, mixture is injected in main line. Pressure sustaining valve opening and closing process is auto adjusted as per main line pressure.

Applications

- Micro-irrigation system for open field fertigation.

- Precision fertigation in protected cultivation (green house/Net house)

- Fertigation for crop in soilless media.

- Hydro phonic nutrient application.

- Aero-phonic nutrient application in protected cultivation.

- With Boom irrigation system.
- To maintain the PH of water for irrigation purpose.
- For Acid treatment.
- Drain water sampling and injection for fertigation.
- R&D plots for fertigation with appropriate quantity.

Achievements

- Fertigation can be done on the basis of EC & PH.
- Precise fertigation can be achieved.
- Manual adjustment of nutrient flow.
- Balancing the PH of irrigation water.
- Frequent operation of solenoid valve.
- Instantly check of EC & PH in sampling cell.
- Automatic operation through Irrigation controller.
- Easy installation.
- Long life with less maintenance.
- High pressure bearing capacity
- Fert-pump multiplication capability.
- Volumetric & sensor based operation through controller.

Installation

Step 1

Unwrap the NutriCare Machine by using sharp knife. Keep knife distant from micro tube fitting, rotameter and hose pipe.

Step 2

Prepare a proper balanced foundation for mounting NutriCare machine. Keep distance minimum 1.5 m and maximum 5 m of NutriCare machine from the main line.

Step 3

Take two connections from main line. These two connections are called inlet and outlet connection port. Reduce the inlet and outlet connection port to 50 mm PVC line.

Distance between inlet and outlet connection port shall not be less than 2 m.

Step4

Connect the inlet connection port of mainline to inlet of NutriCare machine and outlet connection port to the outlet of NutriCare machine. Ensure that all the PVC fitting shall be leak proof. Connection pipe shall be balanced with proper support.

Step 4

All the connected MTA and FTA in inlet and outlet of NutriCare shall be Teflon tight.

Step5

Connect the Sampling tube in main line next to inlet connection port. And the distance between inlet connection port to sample point shall be minimum 1 m.

Step 6

Insert EC & PH probe in sampling cup and tightened the rubber seal and packing.

Step 7

Connect the Hose pipe supplied with NutriCare machine to the inlet of rotameter and other end to the bottom of fertilizer collection tank.

Step 8

Open the electrical control panel and connect three phase power supply to electrical control panel. Connect Controller Digital output to the solenoid valve, and booster pump. Connect analog input signal from EC & PH monitor/ transmitter to the controller.

Step 9

Unscrew the pressure switch and set the pressure minimum 1kg/cm2 with the difference of 0.5 to 0.7 Kg/cm^2.

Step10

Program the controller, start the main pump and set the pressure sustaining valve by rotating pressure pilot nob to achieve the pressure higher than main line pressure.

Step 11

Stop all the system and ensure the leakage. If found then tightened or use Teflon tape at threaded part and Solvent cement at PVC part.

Step 12

Ensure direction of rotation of pump. And finally system is ready to work.

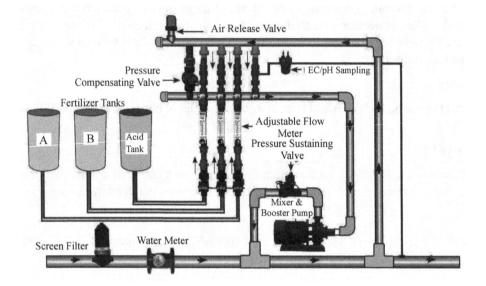

Fig. 5.4: Schematic diagram of positioning fertigation machine in a running drip system \

6

Nutrient Dosage & Fertigation Schedule

There is a lot of confusion and lack of understanding when it comes to scheduling of fertigation. Even crop specialists prepare a fertigation dosage by dividing the total fertilizer quantity by the number of days of crop growth. This is highly erroneous. A scientifically prepared fertigation plan allows coordination of nutrient supply with changing demands of the growing crops which changes with the physiological shifts in the crop. Fertigation planner requires following information to prepare this schedule- which, when and how much plan:

1) Crop growing cycle with development stages and their duration

2) How the crop growth cycle is influenced by local climate.

3) Amount of each nutrient required at each stage

4) Nutrient removal pattern of the crop from the soil/media.

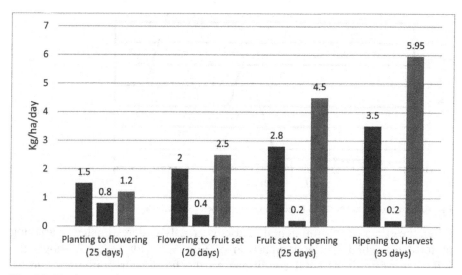

Fig. 6.1: Nutrient uptake rate recommended based on the uptake rate of Field Tomato – (Data *Source*: International Potash Institute (IPI)

Nutrient uptake at any one time in a crop cycle depends on crop characteristics

1) Expected final yield

2) Nutrient content in the harvested part and the residual biomass

3) Environmental conditions- temperature, humidity and light.

4) Soil nutrient status (for soil grown crops)

Considering all these factors specific nutrient recommendations for a crop have to be based on nutrient uptake measurements under conditions as near as possible to those in which the crop is grown. However, this is not always possible especially in the innumerable villages located in large districts of several states of the country. Oftentimes the nearest location from where a recommendation is made can be 400-500 km away from a small farm where the crop is grown.

Therefore it is more or less accepted that generalized fertilizer recommendations can be utilized for fertilization plan for a specific crop and its different cultivars. Nonetheless, fertigation gives a tool to make regional modifications of the fertilizer quanta and timing based on soil and leaf/petiole analysis. A far rapid method of adjustments in the application rate achieving precision is possible.

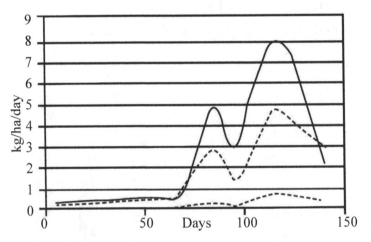

Fig. 6.2: Uptake of nutrients in Green house tomatoes (Source IPI)

Often the effect of environmental factors is grossly underestimated in arriving at fertilizer schedules. A comparison of Fig: 6.1 with Fig: 6.2 shows the influence of environmental conditions on the fertilizer requirement curves. Here the timing and quantities of N, P, K needed for tomato crop are very different in Green house vis a vis open field conditions.

A solution to this issue is use of leaf analysis at various stages of the crop and making real time adjustments in the nutrient application based on the data. Table   illustrates the normal nutrient concentrations in a tomato crop as obtained from leaf analysis.

Table 6.1: Normal nutrient concentrations in trellised tomato (youngest full mature leaves were analysed) Normal nutrient concentrations in trellised Tomato (at 1st mature fruit)

N (% of dry wt.)	2.5-4.0	Cu (ppm in dry matter)	5-10
P	0.3-0.6	Zn	30-40
K	3.0-4.0	Mn	50-100
Ca	0.5-2.0	Fe	100-300
Mg	0.6-1.0	B	30-100
		Mo	0.4

Fertigation schedule normally prepared from the local fertilizer recommendations should be adjusted from time to time by

1. Making regular observation of the crop

2. Identifying any symptoms of nutrient deficiency

3. Analyzing the leaf/petiole and recalibrating the fertilizer quantity and type in the fertilizer program.

4. Adjusting also for any season elongation or change in the developmental phase of the crop.

It is also to be noted that many of the conventional beliefs regarding fertilizer dosage and timing have changed in major crops- Tomato, Capsicum, Banana, Sugarcane after the introduction of fertigation and detailed study of nutrient requirements/removal by crop. This aspect is considered as the collateral technology change associated with the spread of drip technology.

Availability of nutrient ratios with respect to the physiological stage of the crop is essential for arriving at the correct fertigation schedule (see Table below).

Table 6.2: Typical NPK ratios in a fertigation program according to the physiological stage of the crop (Tomato).

Physiological stage	Ratio of nutrients			Observations
	N	P_2O_5	K_2O	
Planting and establishment	1	1	1	High P and K for best rooting and establishment
Vegetative	2	0.5	1.5	High N & K for best plant growth and leaf development
Fruit set –ripening	1	0.3	3	High K for best fruit quality colour, size, flavour)
Ripening- harvest	1	0.1	2	High K for fruit quality aintenance

Source: International Potash Institute (IPI).

Table 6.3: A typical case of arriving at a fertigation schedule is described below.

Crop	Sugarcane
Crop duration	11 months to harvest.
Duration of fertigation	265 days
Recognized growth stages and developmental phases	Seedling phase;
	Tillering Phase;
	Grand growth phase and
	Maturity phase.

Total quantum of fertilizers are determined 1) Using local fertilizer recommendation or 2) by estimating fertilizer requirement for target yield using nutrient removal data available in the literature Scheduling.

Factors considered: total quantum of each fertilizer; Nutrient need (type and quantum) of each physiological stage.

In this case, the fertilizer quanta determined by either one of the above two methods utilized to make the Fertigation schedule. These are then split into many small doses following the uptake curves published after scientific studies (Fig. 6.3). Based on the figure the NPK doses and their ratios are arrived at each stage of cane growth.

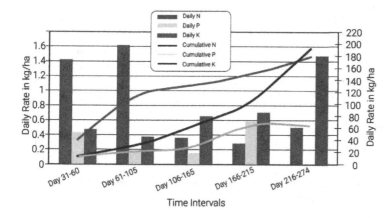

Fig. 6.3: Nutrient uptake curve (NPK only) for sugarcane crop (*Source*: Duclos)

The information in the above figure (Fig 6.3) can be translated into ratios of N P and for each growth period and actual fertilizer doses are prepared based on these ratios.

Because the uptake is physiologically controlled by the crop; the schedule of fertigation also follows the same rates. This way the planter (farmer) is able to match the nutrient demand at different stages in terms of type and quantity.

Table 6.4: The ratios of N and K for sugarcane:

Seedling phase	3 : 1 (N:K)
Tillering phase	1: 1
Grand growth phase	2: 1
Maturity phase	1: 3.5
Seedling phase	3 : 1 (N:K)
Tillering phase	1: 1
Grand growth phase	2: 1
Maturity phase	1: 3.5

A schedule of application on alternate days till 265 days of crop growth as an example is given below. The recommended fertilizer quanta are 119: 69:60 NPK.

Table 6.5: Fertigation schedule for sugarcane

Fertigation is recommended once in two days

S.No	Duration in Days	Dose of Urea once in 2 days kg/ac	Total Urea Doses	Dose of MAP once in 2 days kg/ac	Total MAP oses	Dose of MOP once in 2 days kg/ac	Total MOP doses
1.	15 to 30	5.14	7	0	0	0.4	7
2.	31 - 105	2.75	38	0.74	38	0.4	38
3.	106-207	1.44	50	0	0	0.46	50
4.	207-255	1.2	25	0	0	1.8	25
5.	256-265	0	0	0	0	2.8	5

MAP; Mono Ammonium phosphate
MOP; Muriate of Potash
Source: Soman.P. et al. 2018

Other model schedules

Banana

Major Factors considered;

1) Growth Stages

    i. Early vegetative growth

    ii. Late vegetative growth

    iii. Bunch exit

    iv Bunch and finger growth

2) Fertigation Period 300 plus days.

3) Critical Physiological Stage

Need for K^+ ions for transfer of photosynthate to fruits till the fingers on the last hand is developed.

Table 6.6: Fertigation Schedule for Banana

Total Nutrient Requirement

N-150gm/plant	P-45gm/plant	K-185gm/plant

Total Quantity of Fertilizer Required/acre (Spacing 1.8 x 1.5 m; 1452 pts)

Fertilizer	Quantity (Kg / acre)
19:19:19	192 kgs
0:52:34	59 kgs
13:00:46	458 kgs
Urea	337 kgs

Schedule of Application

Period	Grade	Total Qty (kg)	kg / day / acre
1 - 90 days	19:19:19	192 kgs	2.13
	13:00:46	93 kgs	1.033
	Urea	70 kgs	0.777
91 - 150 days	0:52:34	59 kgs	0.983
	13:00:46	82.5 kgs	1.376
	Urea	117.5 kgs	1.958
151 - 300 days	13:00:46	282.5 kgs	1.884
	Urea	150 kgs	1

Source: Soman.P. Unpublished.

Potato

Factors Considered for Scheduling fertigation of Potato.

1) Growth Stages

   i. Sprouting

  ii. Vegetative growth

 iii. Tuber initiation

 iv. Tuber bulking

  v. Maturation

Table 6.7: Fertigation Schedule for Potato

Total Nutrient Requirement

Schedule of Application		50:25 : 50 NPK /ac	
Period	Grade	Total Qty (kg/ac)	kg /day /ac
At Planting as soil application	DAP	50	
15-30 days	Ammonium sulphate	28.5	1.9
	Ammonium sulphate	71.5	2.9
31-55 days	Urea	31	1.2
	MOP	37.5	1.5
56-70 days	Urea	12.4	0.9
	MOP	21.6	1.5
71-90 days	MOP	24	1.26

Source. Soman .P. unpublished.

Mango (an orchard crop)

i.   Duration and timing of flowering and fruiting.

ii. Duration and timing of active vegetative growth.

iii. Spacing of trees.

Table 6.8: A. Mango Fertigation Schedule for 6 yr Plus Trees at 10 x 10 m Planting

Fertilizer	Quantity	Rate of fertigation	Duration
Urea	1000 g/tree	250 g/tree/wk	June 4th wk – July 3rd wk
	1100 g/tree	275 g/tree/wk	Oct 1st wk – Oct 4th wk
MOP*	800 g/tree	200 g/tree/wk	June 4th wk - July 3rd wk
	870 g/tree	220 g/tree/wk	Oct 1st wk – Oct 4th wk

All 6.25 kg SSP to be applied in three splits, June (2.25 kg /tree). Sept. (2kg/tree), November (2kg/tree) as soil application.
**Apply the SSP in two rings around the tree 1m and 2m away from the trunk, respectively.*
**Place the fertilizer at a depth of 10-15 cm below the soil surface and cover with top soil.*
Source: P.Soman (unpublished)

Table 6.9: B. Mango planted at 3 x 2 m (Ultra High density)

Age of Tree	Fertigation Schedule and Quantity (kg/dose/acre)					
	Month	Number of Doses	Urea	H3PO4	MOP	*MgSO4
1yr	July-Sept	12	1.4	0.5	0.8	0
	Jan-May	20	1.7	0.6	0.9	0
2 yr	July-Sept	12	2.7	1.2	2.3	0.278
	Jan-May	20	1.6	0.7	1.4	0.167
3 yr	15 June-Aug	12	4.5	2.3	3.5	0.555
	Sept	4	1.4	1.2	3.1	0
	Jan-May	20	3.2	1.2	1.5	0.333
4 yr on-wards	15 June-Aug	12	7.2	3.5	4.6	0.833
	Sept	4	2.2	1.7	4.2	0
	Jan-March	12	5.1	1.7	3.2	0.833

Source: U.Chaudhari et al. 2019

7

Fertigation of Green House Crops

The crop growing environment, both soil/media and aerial are controlled in green house. Therefore, the movement of water (liquid and vapour) and minerals (nutrient ions) follow a different pattern altogether. This factor should be considered while preparing the fertigation program. This becomes all the more different if soil-less rooting media is used.

Table 7.1: Differences in Nutrient uptake in open field and Protected (GH) conditions.

Crop	Yield t/ha	Nutrient uptake (kg/ha)		
		N	P	K
Tomato (open field	80	250	35.2	415
Tomato (Protected)	100	200-600	44-88	500-830
Cucumber (open)	60	170	57	224
Cucumber (protected)	300	450-500	88-110	664-830
Melons (open)	20-30	120	11	166
Melons (protected)	30-40	200	22	166

*** Adapted from IFA 1992.*

For green house/shade house crops grown in Soil less culture or in Sand culture fertigation is done by proportional dilution ; and estimated concentration of nutrient is expressed as ppm or kg/m² water. Unlike in the open field soil based cultivation, in soil less media the concentration of nutrients will remain constant in the water throughout the application period; hence it is termed as Proportional dosing of nutrients. In the field soil based cultivation, on the other hand, fertilizer nutrients are applied at certain intervals and their concentration will decrease as a function of time in the solution reaching the root zone.

The growth of vegetables and flowers in greenhouses on sandy media and/or with inert substrates requires a special and precise control of the fertigation, because the CEC of these growing media are very low and therefore they do not provide nutrients. The only source of nutrients is through the fertigation system. Growing plants in containers allows the collection of the leaching water and its comparison with the irrigation water. The measurement of pH, EC and nutrients concentration in the leached solution indicates if fertilizers are being applied in excess or deficiency, and therefore allows the consecutive correction of the fertigation regime. It is recommended to collect the leached solution from the containers and the solution that leaves the drippers, and to compare both solutions on a daily basis. To maintain precision, Automatic computerized devices that measure pH and EC of both solutions and automatically corrects the next irrigation solution according to optimal values entered beforehand.

Fertigation requirements in soilless culture

Fertigation in soilless culture requires high adaptation to the following factors:

1. Plant needs. The ratio between plant nutrient consumption and water needs is a function of crops species and variety, growth period and conditions.

2. Growing conditions (temperature, light, etc.). For example, the mineral absorption of a winter-grown radish is equal to that of a summer-grown crop. However, the water absorption in summer is three to four times greater than in winter.

3. Nutrient solution composition in the roots environment. The pH and EC values of the substrate solution are measured two to three times a week. The solution should be sampled near the dripper and in the drainage. On the basis of these measurements the grower can adjust the addition of NH_4 + to control the pH and the EC of the nutrient solution.

a) Electric Conductivity

A higher value of EC in the leached solution than in the applied solution indicates that the plant absorbs more nutrients than water, therefore we must apply greater amount of water to the plant. On the other hand, if the difference between the EC of the leached solution and the incoming solution is more than 0.4-0.5dS/ m, we must apply a leaching irrigation in order to wash the excess of salts.

b) Chlorides

Sometimes improper management of the irrigation regime may lead to an unwanted accumulation of Cl ions present in the irrigation water. If the Cl concentration in the leachate is higher than the Cl concentration in the incoming solution and surpasses 50mg/L, it indicates a chloride accumulation in the root zone. Then it is recommended to apply an irrigation without fertilizers to leach the chlorides.

c) pH

The optimal pH value of the irrigation solution must be around 6 and the pH of the leaching solution should not exceed 8.5. A more alkaline pH in the leaching water indicates that pH in the root zone reaches a value that causes phosphorus precipitation and decreases micronutrients availability. When pH in the leachate is higher than 8.5, we must adjust the NH_4/NO_3 ratio of the irrigation solution by increasing slightly the NH_4 proportion. When pH in the irrigation solution is higher than 6, we must inject acid to the solution to lower the pH.

An example for Tomato is given below. Similar programs need to be prepared for each crop.

Table 7.2: Nutrient Concentrations in the fertigation solution at different stages of Protected tomato crop

Nutrient Concentration in the Dripper Solution for Tomato in Green House					
Growth Stage	Nutrient quantity (g/m³) in the solution				
	N	P	K	Ca	Mg
Planting- Vegetative	164	48	174	100	50
Flowering- Fruitset	190	48	250	100	50
Ripening -1st Harvest	200	48	280	100	50
1st harvest- End of crop	205	48	250	100	50

Source: Literature of Haifa Fertilizers

In practice, however Stock solutions of appropriate nutrients were prepared and kept in Tanks. These were then injected at predetermined rate into the fertigation system as per the program. For example, for the Fertigation Program the following Stocks are kept.

Table 7.3: Nutrient Stock solution for Fertigation of Tomato (Protected) in Tank A and B. Stock Solutions and Their Composition

Growth Stage	TANK A		TANK B	
	g/liter			
	Pot.Nitrate (Multi-K)	Calcium Nitrate	MAP	MgSO$_4$
Planting- Vegetative	91	107	36	150
Flowering- Fruitset	130	107	36	150
Ripening -1st Harvest	147	107	36	240
1st harvest- End of crop	100	107	36	120

Injection rate of the stock solution is 5 liter/m³.   Source: Haifa Fertilizers

The total system of fertigation can be automated and a NutriCare can carry out the fertigation program without human error.

8

Fertigation Management- Future

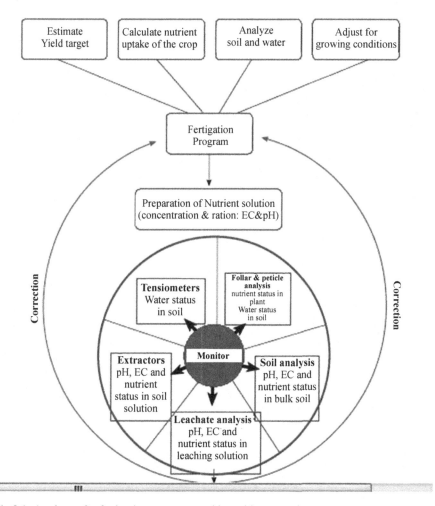

Fig 8.1: A scheme for fertigation program working with automation

In modern farming with fertigation machines (NutriCare) and automated drip irrigation systems, fertigation set up schemes like the above are put into practice. This scheme will stream line the installation, operation and monitoring of the entire process. An informed grower would be able to control crop performance more precisely and achieve his objectives.

Another area where developments are expected is in the sensor technology that in conjunction with automated control would further improve precision of application of nutrients. However, soil nutrient sensors are still a long way to find. Currently, sensors are available only for heat (temperature), water (soil moisture) of the soil. Mineral nutrient state of soil is yet to be monitored by sensors. If and when these sensors are made available we would attain the best possible precision in farming. Fertigation then would be more accurately done.

References

1. Kafkafi. U and J. Tarchitsky. 2011 Fertigation –A tool for efficient fertilizer and water management. International Fertilizer Industry Association (IFA) and International Potash Institute (IPI). Paris, France.
2. Mengel .K, and E.A.Kirkby, 2001. Principles of Plant Nutrition. 5th edition. Dordrecht: Kluwer Academic Publishers.
3. Ankerman. B.S, and Large .R., (Ed.) 2005 Soil and Plant analysis Agronomy Handbook. A&L Agricultural Laboratories.
4. Soils Bulletin 1980. Soils and Plant testing as a basis of Fertilizer recommendations. FAO bulletins. FAO portal.
5. Zende, G.K., 1984. Efficient use of fertilizers for increasing sugarcane productivity. 34th Convention of DSTA, Pune, India.
6. Hoagland, D.R. and D.I. Arnon, 1950. The water culture method for growing plants without soil. Circular, California Agriculture Experiment Station. Vol. 347 (2nd Ed.).
7. International Potash Institute (IPI) 2005. Fertigation Proceedings: Selected papers on the IPI-NATESC-CAU-CAAS International Symposium on Fertigation, (ed. P. Imas and M.R. Price), Beijing, China, 20-24 September 2005.
8. Duclos International, www.duclosinternatioanl.eu
9. Soman P., Singh. S, Balasubramaniyam, V.R. and Velmurugan, R., 2018. Differential Application of Fertigation levels to Enhance Cane yield to Maximum of Sugarcane Cv.CO 86032 under Subsurface Drip Irrigation. International Journal of Agriculture Sciences, ISSN: 0975-3710 & E-ISSN: 0975-9107, Volume 10, Issue 11, pp.- 7104-7107.
10. Chaudhari A.U., B. Krishna, P.Soman, V.R. Balasubramanyam, (2019) Development of A Package for Intensive Cultivation of Mango Using Ultra-High-Density Planting (UHDP), Drip and Fertigation Technologies for Higher Productivity. International Journal of Agriculture Sciences, ISSN: 0975-3710 & E-ISSN: 0975-9107, Volume 11, Issue 23, pp.- 9200-9200.
11. International Fertilizer Association, 1992. IFA Technical Conference. The Hague, Netherlands, 6-8 October , 1992.
12. Haifa Fertilizers- Tomato crop guide: Dynamics of Nutritional requirements.

Appendix

Conversion Tables

Table 1: Areas and weights, yields and application rates

From	To	Multiply by	From	To	Multiply by
Acre	hectare	0.405	Hectare	acre	2.471
Kilogram	pounds	2.205	Pound	kilogram	0.453
Gram	ounces	0.035	Ounce	gram	28.35
Short ton	Metric ton (MT)	0.907	Metric ton (MT)	Short ton	1.1
Gallon (U.S)	liters	3.785	Liter	gallon	0.26
kg/ha	pound/acre	0.892	kg/ha	pound/acre	1.12
MT/ha	pound/acre	892	pound/acre	MT/ha	0.001

Table 2: Elements and their oxides

Conversion factor from elements to their oxides			Conversion factors from oxides to their elements		
From	To	Multiply by	From	To	Multiply by
N	NO_3^-	4.43	NO_3^-	N	0.23
N	NH_4^+	1.28	NH_4^+	N	0.82
P	P_2O_5	2.29	P_2O_5	P	0.44
K	K_2O	1.2	K_2O	K	0.83
Ca	Ca	O 1.4	CaO	Ca	0.71
Mg	MgO	1.66	MgO	Mg	0.6
S	SO_3	2.5	SO_3	S	0.4
S	SO_4	3	SO_4	S	0.33

Table 3: Atomic and equivalent weights

Element	Atomic Weight	Valence	Equivalent weight
H	1	1	1
N	14	1	14
K	39.1	1	39.1
Ca	40	2	20

	Mg	24.3	2	12.15

Table 4: Soil Nutrient Ratings of Indian Soil

S.No.	Parameters/ Nutrients		Rating		
1.	pH		Acidic	Neutral	Alkaline
			<6.5	6.5-7.5	>7.5
2.	Electrical Conductivity		Safe	Moderate	Unsafe
		(dS/m)	<1.0	1.0-2.0	>2.0
	OC and Nutrients	Low	Medium	High	
3.	Organic Carbon (OC)	(%)	<0.5	0.5-0.75	>0.75
4.	Available Nitrogen (N)	kg/ha	<280	280-560	>560
5.	Available Phosphorus (P2O5)		<23	23-56	>56
6.	Available Potassium (K2O)		<130	130-336	>336

Table 5: Sufficiency range of nutrients concentration in index tissue (leaf or Petiole) of major crops

Nutrient	Mango	Banana	Grapes	Orange	Pomegranate	Apple	Coconut	Potato	Onion	Tomato	Cucumber	Sugarcane
N, %	1.0-1.5	2.5-3.0	1.32-2.21	2.20-3.5	0.91-1.66	1.90-2.6	1.8-2.1	3.30-4.5	5.0-6.0	4.0-6.0	4.50-6.0	2.0-2.60
P,%	0.08-0.25	0.18-0.40	0.38-0.75	0.13-0.5	0.12-0.18	0.14-0.4	0.11-0.12	0.23-0.5	0.35-0.5	0.25-0.75	0.34-1.25	0.18-0.30
K,%	0.40-0.90	2.3-4.0	1.14-2.20	1.20-3.0	0.61-1.59	1.50-2.0	1.2-1.4	3.10-4.5	4.0-5.5	2.9-5.0	3.90-5.0	1.10-1.80
Ca, %	2.0-5.0	0.70-1.4	0.74-1.14	1.10-4.0	0.77-2.00	1.20-1.60	0.35-0.50	0.70-1.2	1.0-2.0	1.0-3.0	1.40-3.5	0.20-0.50
Mg, %	0.20-0.50	0.25-0.40	0.63-1.10	0.30-0.50	0.16-0.42	0.25-0.4	0.25-0.35	0.35-1.0	0.25-0.40	0.4-0.6	0.31-1.0	0.10-0.35
S, %	-	0.26-0.5	0.14-0.27		0.16-0.26	0.20-0.40	0.15-0.20	-	0.5-1.0	0.4-1.2	0.40-0.7	-
Fe, ppm	50-250	100-300	54-80	60-150	71-214	50-300	40-115	40-100	60-300	40-200	50-300	40-250
Mn, ppm	50-250	200-2000	76-174	25-200	29-89	25-200	60-120	40-250	50-250	40-250	50-300	25-400
Zn, ppm	20-200	13-50	53-132	25-200	14-72	20-100	60.00	20-50	25-10	20-50	25-100	20-100
Cu, ppm	7-50	6-30	5-10	6-100	29-72	6-50	12-13	-	15-35	5-20	7-20	5-15
B, ppm	-	-	-	25-100	-	25-50	8.00	25-75	22-60	25-60	25-60	4-30
Mo, ppm	-	-	-		-	-	-	-	-	-	0.8-3.3	0.05-4.0

Table 6: Fertilizer calculation table for different concentration solutions of N, P and K

S. No.	Soluble Fertilizers	Nutrient Content (%)		Quantity of Fertilizer Required for Different Nutrient Concentration								
				PPM					%			
		Nutrient	%	1	10	50	100	1000	1	2	5	10
				Quantity (mg/lit)					Quantity (g/lit)			
1.	Urea	N	46	2	22	109	217	2174	22	43	109	217
2.	Single Super Phosphate (SSP)	P	16	6	63	313	625	6250	63	125	313	625
3.	Orthophosphoric Acid (H3PO4)	P	73	1	14	68	137	1370	14	27	68	137
4.	Potassium Chloride (MOP)	K	60	2	17	83	167	1667	17	33	83	167
5.	Potassium Sulphate (SOP)	K	50	2	20	100	200	2000	20	40	100	200
6.	Mono- Ammonium Phosphate MAP)	N	12	8	83	417	833	8333	83	167	417	833
		P	61	2	16	82	164	1639	16	33	82	164
7.	Di- Ammonium Phosphate (DAP)	N	18	6	56	278	556	5556	56	111	278	556
		P	46	2	22	109	217	2174	22	43	109	217
8.	Mono potassium Phosphate	P	52	2	19	96	192	1923	19	38	96	192
		K	34	3	29	147	294	2941	29	59	147	294
10.	Potassium Nitrate	N	13	8	77	385	769	7692	77	154	385	769
		K	45	2	22	111	222	2222	22	44	111	222
11.	Complex fertilizer 19:19:19	N	19	5	53	263	526	5263	53	105	263	526
		P	19	5	53	263	526	5263	53	105	263	526
		K	19	5	53	263	526	5263	53	105	263	526
12.	Complex fertilizer 12:32:16	N	12	8	83	417	833	8333	83	167	417	833
		P	32	3	31	156	313	3125	31	63	156	313
		K	16	6	63	313	625	6250	63	125	313	625
13.	Complex fertilizer 10:26:26	N	10	10	100	500	1000	10000	100	200	500	1000
		P	26	4	38	192	385	3846	38	77	192	385
		K	26	4	38	192	385	3846	38	77	192	385
14.	Complex fertilizer 15;15;15	N	15	7	67	333	667	6667	67	133	333	667
		P	15	7	67	333	667	6667	67	133	333	667
		K	15	7	67	333	667	6667	67	133	333	667

Table 7: Fertilizer calculation table for different concentration of solutions of Micronutrients .

S. No.	Soluble Fertilizers	Nutrient contents (%)							Quantity of Fertilizer Required for Different Nutrient Concentration						
		Mg	B	Fe	Mn	Zn	Cu	S	PPM				%		
									Quantity (mg/lit)				Quantity (g/lit)		
									1	10	100	500	0.1	0.5	1
1.	MgSO4	16	-	-	-	-	-	13	6	63	625	3	6	31	63
2.	Borax	-	11	-	-	-	-	-	9	91	909	5	9	45	91
3.	FeSO4	-	-	19	-	-	-	12	5	53	526	3	5	26	53
4.	MnSO4	-	-	-	31	-	-	15	3	32	323	2	3	16	32
5.	ZnSO4	-	-	-	-	21	-	13	5	48	476	2	5	24	48
6.	CuSO4	-	-	-	-	-	24	13	4	42	417	2	4	21	42